# Accidenta Cowgirl

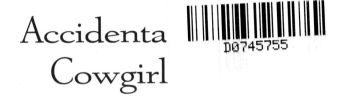
D0745755

## Six Cows, No Horse and No Clue

# Accidental Cowgirl

## Six Cows, No Horse and No Clue

Mary Lynn Archibald

CLOUD LAKE PUBLISHING
Healdsburg, California

979.414
Archibald

Manufactured in the United States of America.

Cloud Lake Publications
1083 Vine Street, #185
Healdsburg, CA 95448
U.S.A.

www.winecountrywriter.com

Accidental Cowgirl: Six Cows, No Horse and No Clue/
Mary Lynn Archibald
-1st ed.

"…*In masks outrageous and austere*
*The years go by in single file;*
*But none has merited my fear,*
*And none has quite escaped my smile.*"

—Elinor Wylie

"...*as humans we require support for our spirits, and this is what certain kinds of places provide. The catalyst that converts any physical location-any if you will-into a place, is the process of experiencing deeply. A place is a piece of the whole environment that has been claimed by feelings.*"
—Alan Gussow

"*Different places on the face of the earth have a different vital effluence, different vibration, different chemical exhalation, different polarity with different stars; call it what you like, but the spirit of place is a great reality.*"
—D.H. Lawrence

"*Standing on the bare ground...all mean egotism vanishes. I become a transparent eyeball; I am nothing; I see all; the currents of the Universal Being circulate through me; I am part and particle of God.*"
—Ralph Waldo Emerson

*For Carl,*
*the love of my life*

# Contents

# Introduction

*"To live more simply is to unburden ourselves—to live more lightly, cleanly, aerodynamically."*
—Duane Elgin, *Voluntary Simplicity*

SIMPLICITY. Lead me to it, you say from the midst of your overburdened, overscheduled, list-making life. If this is your desire, you are not alone. Not so fast. Before you sink your savings into that country squire lifestyle and all that goes with it, perhaps you should take a cold shower and read this book.

We lived that life for 12 years, and let me tell you right up front: the simple life isn't really that simple, and despite my persistent romantic fantasies, I rarely managed to look like I'd just emerged from the pages of an L.L. Bean catalog. (Those people never get muddy, and they wear bras in town).

\* \* \*

In 1990 we heard the wilderness call to us and, God help us, we answered.

Part of the problem was that we had no business trying to run a full-time ranch with no experience. People tried to tell us that, but we weren't listening.

Each time Carl and I left for the ranch, we deluded ourselves that as soon as we arrived, we could relax, hike, swim, read, paint and picnic on our own land, and in general, enjoy the place.

But each time we arrived, there was a fussy water system to deal with, there was wood to haul and a fire to make, there were cows to care for, plants to water, fields to mow, weeds to pull, a house to clean, barns to repair, critters to eject from our living space, ditches to dig, roads to fix, rampant vines to vanquish. By each bedtime, we had not read a chapter, and if we hiked, we carried packs full of tools and barbed wire.

\* \* \*

And we were not alone in such a wild endeavor. In 1993, I began noticing a trend that has continued and strengthened up to the present day.

As cities are becoming more crowded and the quality of life therein more problematic, and as technology becomes ever more fine-tuned, people can choose to work most anywhere they like and skip the office.

As people go farther afield to find quiet and peace, even the suburbs are not acceptable to some, so country living can seem like the answer for more and more people.

That's no doubt one reason. Another is to live more authentically, closer to the land, closer to family.

—Some examples:

The September 6, 1993 issue of *Time* magazine featured a cover titled, **"Boom Time in the Rockies: More Jobs and Fewer Hassles Have Americans Heading for the Hills."** The cover story, "Sky's the Limit," said, "The home of cowboys and lumberjacks has become a magnet for Lone Eagle telecommuters and Range Rover-driving yuppies. So far, it's been a booming good time…" that combines "the yearning for a simpler, rooted, front porch way of life, with the urban bred, high-tech worldliness of computers and modems."

I saved that issue.

Then the March/April issue of *Motorland* magazine (now *VIA*), featured prominently an article by Lynn Ferrin, called **"Ridin' the Range Once More—don't we all, deep down, want to be cowgirls and cowboys?"** In it she samples cattle herding with a group of greenhorns, led by a trail boss and 10 or so cowboys, as they experience driving 300 head of cattle 60 miles to Reno for the big annual rodeo.

And just as they did in the movie, "City Slickers," folks with the yen to experience the Old West actually *paid* to do this.

I saved that one, but I don't recommend that you sign up for a trail ride until you've read this book. By the time I got around to reading all those rhapsodic articles, it was too late for me.

I was already herding obstinate cattle and substitute-teaching at the two-room school by then—herding obstinate children and trying vainly to explain Renoir to kids who were more conversant with the name Remington (Firearms).

There were no high-tech jobs in our area.

There were no jobs—the listing of the Northern Spotted Owl and the Marbled Murrelet as endangered species having effectively shut down the logging industry and the businesses that depended on loggers with money in their pockets.

This was the heyday of Julia "Butterfly" Hill and her tree-sitting media event; of protesters who lay down in front of bulldozers to keep old-growth redwoods intact.

It was the era of CAMP (the Campaign Against Marijuana Production), whose helicopters regularly hovered low to scan our mountaintop for illicit crops, whose growers were unfortunately one of the few prosperous groups in the whole of what we knew as "The Emerald Triangle."

Challenging times in the wilderness.

Earlier residents had tried farming and sheep ranching, but the deer ate the crops, cougars and bears ate the deer, and coyotes preyed on the sheep.

By the time we arrived, cattle-ranching was the principal method for earning enough money to keep the roof from leaking and the family fed.

But, of course, farming and ranching are always problematic, dependent as they are on weather and the vagaries of nature.

At Twin Creeks Ranch, we had beautiful countryside, some of it level, but just not enough of it to support a large herd.

And a large herd, unfortunately, was exactly what was needed to make enough money to pay the property taxes and the vet bills, and to buy the supplemental feed that was always needed before the green spring grass came back each year.

The old-timers knew this of course, and undoubtedly had a few laughs at our expense, watching us struggle with our small herd on our small acreage—120 acres, after all, was not a big spread in those parts, and most ranchers we knew had more like 1,000 acres, or else they leased someone else's land on which to graze their cattle when it came time to drive them out of the national forests.

At first, grazing our cattle was not a problem for us because we only had six cows, but in fewer years than we expected, our herd had grown to 26, and showed no sign of slackening its reproductive pace.

And quite unexpectedly, the sweet things became very hard for us to part with, and we would keep them longer than we should have before they were sold. The other factor, it soon became clear to us, was the problem of the wandering bulls.

Our cows were in the habit of coming into estrus all at the same time (a not uncommon occurrence with such creatures), thus amplifying the power of the pheromones they were sending out into the ether around our ranch. It was pretty damned potent.

So greatly amplified was it in fact, that we never

needed to own a bull. Even though Carl felt compelled one year to run an ad in the local newsletter, entitled "No bull!" we generally had the opposite problem, for bulls inevitably found us. Often, it was difficult to tell just *whose* bulls they were. Nevertheless, they would manage to squeeze over or under or around or through our fences to get to our succulent herd of single cows.

We would simply wake up one morning and find them cavorting shamelessly in the field with some randy fellow who showed no inclination to head back home anytime soon.

"Oh, those hussies!" Carl would say, chuckling and rubbing his hands together briskly.

And so, our little herd grew into a big herd—for us, at least—too big for 120 acres to sustain.

I tried farming for a while. Carl laid out what seemed like miles of irrigation pipe so I could grow herbs in quantity, but that, too, required larger acreage and more help than I could afford, to assist me in working the fields.

Anyway, after repairing the barns each spring, bucking hay into them, cutting and splitting four cords of wood for winter, planting thousands of seeds and erecting nearly two acres of deer fencing (over which deer were seen to sail effortlessly the day after its completion), we finally accepted that we only had enough energy between us to take care of the cattle, find and fix the leaks in both the house roof and the water lines, and grow enough vegetables to keep our big chest freezer well-stocked nearly year-round.

I think we swam in our lake a total of 12 times in 12 years, but we did manage to cook some fabulous meals with our home-grown, fresh veggies, sit on our deck in the cool of quite a few summer evenings, and ease into a good many hot baths after our days laboring in the fields. Though our bones ached, we were thankful.

* * *

Still harbor a romantic image of the bucolic life style? Here are a few questions for you to consider:

Are you fit? You won't need your stairclimber where you're going.

Do you want adventure? How straight can you shoot?

You DO know how to shoot a gun, don't you?

You may not have to deal with teenage gangs, but you might have to defend your territory against marauding cattle, dogs, mountain lions and heavily armed marijuana farmers.

At the time of our residence in Zenia, there were two lawmen—one sheriff and one deputy—for an entire county that measures 3,208 square miles and consists mainly of three national forests—Shasta-Trinity, Six Rivers (which bordered our ranch on one corner), and Mendocino National Forest, which meant you were mostly on your own when it came to law enforcement and property rights.

The sheriff's office was in Weaverville, much closer to I-5 than we were, over mountain roads that could, without notice, become impassable due to fallen trees, road work, mud or rock slides or collisions of logging trucks with deer or cattle.

This meant that you never made the trip (which under normal conditions and depending on bravery or fool-hardiness, took from two to four hours), without sufficient water, gasoline, food and blankets enough for a protracted siege.

Are you adaptable? I went from a 9-to-5 scheduled day, to a non-scheduled day that began at dawn, and was full of the aforementioned chores. Such days generally include a lot of open-ended visiting—which can kill the better part of an afternoon—and end when you decide to forgo that hot bath and just fall into bed, say at sundown.

No TV, no movies, no AM radio, and only two FM stations—your choice between Country-Western and New Age-old hippie.

You're going to need a freezer for all that fresh produce you'll be growing, because last time I checked, all that was available at the general store was iceberg lettuce, yellow onions, cold-storage apples, oranges and bananas and the occasional unripe tomato. Oh, and of course the staple of choice in the mountains—beer.

And you might need a heavy-duty canner to preserve the orchard's bounty—that is, whatever is left after the deer have jumped your deer fence and feasted on your prize apples, peaches and plums. (The birds ate the cherries, though one year they thoughtfully left us enough for a single cherry pie.)

And don't forget your sunscreen, mosquito repellant, snake bite kit, an assortment of big sunhats (in the mountains, you burn much faster), your poison oak medicine, your bee sting injection kit, your asthma and hay fever medicine, and flea and tick repellant for yourself and the dog.

Don't develop any severe illness or sudden health emergency. The nearest medical help is available by telephone of course, if you have one...

It's best to be healthy going in, because the nearest hospital is either 30 bumpy miles down the mountain, which takes about an hour, or 30 minutes by life-flight helicopter to you (they can land in your pasture, but you'll need to put up a windsock so they can find you), and then it takes the other 30 minutes to get back to the hospital in Redding by helicopter—so it's an hour either way.

And don't break a leg while you're out of earshot of the house—your spouse won't miss you till you fail to show up for supper. Don't count on the dog to find you.

Oh, and you'll need a machete and a cordless, heavy-duty brush cutter or bulldozer to clear out that poison oak and the blackberry vines, a tractor to mow the grass around the house (it's a fire hazard if you don't), and maybe a log-splitter and a gas-operated chainsaw, because firewood is expensive and your woodstove will likely be your only source of heat.

\* \* \*

When we were the stewards of the ranch (from 1990 to 2002), many people back in the hills had no telephones, so the folks at the general store in Kettenpom took messages for them and posted them on the wall behind the cash register. The blessedly phone-less would saunter in to buy provisions and to check their messages from the outside world once a week.

If they didn't show up at the store during that time for messages and groceries, it was understood that someone would be dispatched forthwith to look for them.

Thus, fortunately for her, when my painter friend, Sylvia, fell off her roof while filling bird feeders during one weeks-long interval between her husband's visits from the civilized world, someone noticed she hadn't been in lately for her messages or to renew her supply of sherry, (a vice we cheerfully shared), and went to check on her.

She was amazingly unhurt, if you didn't count her dignity.

\* \* \*

Still interested? How easily can you separate yourself from your workaday life, now that it's common to carry videophones, pagers and hand held computers everywhere you go?

By the way, what do you know about cows?

More than we did, I hope.

\* \* \* \* \*

# Acknowledgements

I've so many people to thank. First and foremost is my husband and partner of 23 years, Carl Sutter, for enduring the countless readings and re-readings of this book, for enduring my occasional bouts of madness during the writing of it, and for making the book possible in the first place, by following his dream, all of which made it possible for me to follow mine...

Without Twin Creeks Ranch, there'd have been nothing quite so challenging to write about, and though it encumbered our lives at a time when we were crazy to take on yet another project, it was, and remains, one of the happiest times in my life.

Moira Reynolds Bessette, shamefully unacknowledged in my previous book, *Briarhopper*, has been my editor for several years now, and keeps me on track and honest. She has been an invaluable inspiration for my career.

Robert Aulicino is responsible for my wonderful cover design and the interior of this book.

Jeri White was our peerless ranch manager, of course, who by her very presence enriched our lives, and continues to do so.

Necia Liles (an author herself) graciously took up the slack when I bowed out of Friends of the Library board membership and newsletter duties in order to finish this book. Her keen eye for editing helps me regularly.

Kathleen Murry, Esq. is a longtime friend/attorney who has dispensed legal advice and regular ego boosts, ever since we met in a networking group more than 25 years ago.

My wonderful children, Jamie and Miles TeSelle, I want to thank for their belief in me, for their moral support and their love, and for their help punching cows. I must also thank Miles' wife, Holly Hood, whose veterinary knowledge and love of animals helped us out of a few tight spots.

Carl's wonderful children, Amyre Ready and Tim Sutter have always been supportive of my writing, and Tim had a big role in helping both Carl and me at the ranch.

Last, I want to thank Marvin and Nancy Scott, who also loved and worked amazingly hard on the ranch, and who remained fast friends and advisors throughout our Twin Creeks experience right up to the present day, and who left us with such a priceless legacy.

Thank you all, for everything.

—MARY LYNN
  April 2007

P.S.: Some names have been changed to protect the innocent (or perhaps, even the guilty).

CHAPTER 1

# Land

*"Life is a sum of all your choices."*
—Albert Camus, 1957

*"You can't always get what you want."*
—Mick Jagger

MICK WAS RIGHT ABOUT ONE THING. BUT, AS WE discovered by doing way too many things the hard way, you can get what you need. Eventually. What we needed, Carl and I, were like so many other things in our relationship: two vastly different things; yet our 12-year experiment in country living brought us closer in unexpected, almost magical ways that we were only able to appreciate long after our rural adventure had ended.

It had started, of course, as a dream. Carl's dream, not mine.

Sure, at 55 Carl was already contemplating retirement from his landlord business, but meanwhile he was looking for a place to hide.

A native of New York City, Carl came to California in search of space. Land. "I am a peasant at heart," he declared. "I need to find a quiet, rustic retreat where I can relax, contemplate nature and ponder life's big questions."

Big deal, I thought. I'd grown up on an apple and chicken ranch in the tiny town of Soquel, California, where my dad was assistant postmaster and we knew everybody in town—a circumstance I did not consider desirable.

Nevertheless, Twin Creeks Ranch was, at least in concept, just the place Carl's mind had conjured. But it soon became much more. *Way* too much more. "I need some sort of big spread," said he. "Acreage. And water. Preferably in motion. And rocks. Big rocks. Lotsa big rocks." My own list (at 54) was more modest: Armfuls of wildflowers in the spring, room to grow a few vegetables, and a small but elegant country house I could redecorate with antiques and drape with riotous chintz.

My nearly 15 years as an interior designer had given me very big ideas. Also, I must confess here that early access to too many issues of Town and Country magazine had led me to picture myself as one of those anorexic, Patrician women they feature, posed feeding her thoroughbred horses in front of the country estate, which would be elegant but barely visible in the vast, green distance.

She is beautifully but practically coiffed and wearing the latest Ralph Lauren barn jacket over custom-made designer jeans; her English rubber boots unsullied by mud or manure; extending to her horse a red apple that perfectly matches her carefully manicured, unchipped nails. My God, I thought. I could be that woman!

It was for this I had made the laborious climb (chipping my own nails in the process), from country to suburbs to city, acquiring an admittedly thin shell of sophistication along the way.

No matter. I had done a little acting in my youth. (I wouldn't mention my chorus girl days, of course.) I was *made* for this role.

\* \* \*

We had met, Carl and I, through a newspaper ad in the spring of 1984. He was trim, dark-haired and jauntily mustached. His ad in the Santa Rosa Press Democrat read as follows:

COMPLEX SWM, 47, passes for 39, going on 18, handsome, ex comm'l banker, clergyman, dream analyst, now building own home. Listens to blues, gardens, backpacks, stops for garage sales, writes poetry, reads Countryside (magazine). Immersed in Jungian psych., strong spiritual side but no kook. Well educ. and traveled. Full of exper., innocence, wonderment, entrepreneurial, romantic, down to earth, whimsical, shy, outgoing, playfully comepetitive, great companion, friend, confidante. SF, please have similar qualities and hammer.

To which I replied (I think):

"I am a tall, willowy green-eyed blonde who owns a hammer. Interested? Call this number…" and so on. (I actually have hazel eyes, but I thought green sounded better. The rest was true, except for the fact that I'd never heard of Countryside magazine. He's never let me forget it.)

It was the first ad I'd answered. And he didn't lie—except for the part about being "playfully competitive." This man is *seriously* competitive. But then, so am I.

Not surprisingly, he had a lot of replies to his ad.

We met in a Chevy's restaurant bar. My fault. Very noisy. We talked for a couple of hours, and I was favorably impressed. He had all his own hair and teeth—a plus in the over-40 dating market.

We decided to meet again.

That first meeting was the event that had made us a couple for nearly five years when we acquired the ranch—and it was a newspaper ad that was about to change our lives again.

We started looking for country property in the fall of 1989, barely one year after moving into a half-finished house that Carl had been building with his own hands for the previous seven. But that was a different dream. The plan to further encumber himself at that stage was probably not too practical, but like so many things in life it had less to do with logic than with subterranean urges that would only reveal themselves with time.

\* \* \*

When we finally found our Shangri-la, we weren't even looking. We had already spent six months searching for land, and nothing we had seen in that time had been even close to the shimmering vision we thought we shared.

Garberville was merely a rest stop on a weekend getaway trip. It was late spring, and the tiny town was full of tourists on their way to see the giant redwoods or fish the Eel River or camp, or just hang out and observe the colorful locals in the time warp that was Garberville.

Depending on the season, there is plenty to do around Garberville. There are the Shakespeare and jazz festivals for the highbrow crowd, and the Redwood Run (annual gathering of Harley-Davidson devotees) for those who love the open road, and Reggae on the River for the counterculture wannabes.

But it wasn't the festival season, we didn't own a Harley, and since we had no desire to relive the 'Sixties, the people watching soon palled.

By Sunday morning we were ready to leave, but over breakfast at the Woodrose Café, as Carl, out of long habit, was reading the real estate section of the local paper, he suddenly kicked me under the table and said, "Listen to this!"

"Yeah, okay, what?" I said, yawning and taking another sip of lukewarm coffee.

> Undoubtedly one of the most beautiful properties in California. 120 acres of lush, verdant meadow; fenced pasture; groves of fir, pine and cedar, as well as madrone, buckeye and alder; a one-acre pond and two all-year streams with rushing cataracts, falls and shaded pools.

The ad went on to describe the panoramic vistas to be seen from the rooms and decks of the spacious residence as well as the two-story guesthouse, the fruit orchard; the deer-fenced vegetable garden; the three barns and complete setup for horse and cattle ranching; and the southwestern exposure, both above the fog and below the snow.

He looked at me over the paper. "Whaddaya think, huh? Huh?"

An instinctive hunter, Carl is never more excited than when he is tracking down a deal, especially a real estate deal. I could tell he was on the scent now; reason replaced by sheer animal joy.

He rose.

"I'm gonna call the Realtor. See if he can show us the place today."

"Today?"

"Come on, we're here. We might as well take a look."

"Okay," I said, but I don't wanna get home late. I've got things to do."

\* \* \*

Carl emerged from the phone booth two minutes later, smiling broadly.

"He'll meet us at the Zenia General Store in an hour."

"Great. Where the hell is Zenia? How soon should we leave?"

"Right now. He said it takes an hour to get there."

"Really?"

The first thing we noticed on the drive up was the bullet holes.

Every road sign that we passed had a pattern of them. Trouble in paradise.

\* \* \*

Tourists at Zenia General Store

A scenic, bone-rattling hour later, we pulled up in front of a dilapidated white wooden building with a broad, sagging front porch, which bore a sign proclaiming it to be the Zenia General Store. I couldn't help but notice that there were no other storefronts in sight.

It was, as it turned out, the entire town.

On the faded porch in an ancient aluminum patio chair, sat a leathery old man who regarded us with mild curiosity.

"Ah, we're looking for Mike," Carl said, speaking slowly and loudly, in case the old man was deaf.

Jerking a thumb over his right shoulder, he indicated the door.

"In there."

In the dim light the large, cool, musty room revealed itself. On our left were three enamel and glass deli cases, all empty, and over them hung three huge paper wasp nests. On the back wall were black-and-white photographs of Zenia buried in two feet of snow.

The groceries were stacked on the dark rear wall of the room. There wasn't much available: a few cans of pork and beans, tomato sauce and chili, a few bags of potato chips, and an awful lot of six-packs of beer. An old woman sat huddled by the woodstove in a wheelchair, vainly attempting to warm herself.

The right front corner of the room was completely whitewashed, and over a small, shuttered window was a sign that read, "Zenia Post Office." Next to that was a tall, skinny row of shelves, which bore a few tattered paperback romances—the Zenia Lending Library.

Out of the darkness came a youngish, slender man in a down jacket, followed by a very fit older man in jeans and a crisply pressed plaid shirt, well-worn boots, and a straw cowboy hat.

"You must be Carl. I'm Mike," said the younger man. This is Marvin Scott, the owner of Twin Creeks Ranch. Glad you found us."

Already we were out of our depth. Meeting the property owner was unexpected, as in most real estate deals the tactic is to keep buyer and seller as far apart as possible.

But Marvin, tanned and fit at 72, was not about to be left out. We shook hands. Marvin had a strong, callused grip. We liked him at once.

One more mile of steep, dusty, rutted road and we were at the gate to Twin Creeks Ranch, from which we could see more mountains to the west, and beyond, the King Range, fringed with wispy coastal fog.

Stopping at the ranch house to meet Nancy, Marvin's wife, we parked the cars and started out on foot to look at the land.

Carl wanted to see all of it, and Marvin couldn't wait to show it to him. The two of them charged ahead like old friends, while Mike and I (the two youngest members of the party) puffed along behind.

After an hour of nonstop rambling, Marvin was still as excited as a 10-year-old with a new go-cart, showing us where his neighbor had gotten his bulldozer stuck in the mud while digging the pond, and where he'd had to shoot his old cow, and where a deer hunter had shot his steer, apparently mistaking it for a buck.

By that time I had a migraine and Mike was winded, so the two of us went back to the house to rest and talk to Nancy. She gave me a tour of the main house, the orchard, and her flower garden.

Then she seated me facing the long windows that looked out on the view down the valley to the western mountains. I still don't remember much of what she said, so taken was I by the green vastness and majesty of the place.

The cattle had cropped the tender spring grass so close it resembled a golf course with barns.

When Carl and Marvin returned, I could tell from the stunned expression on my sweetheart's face that our fate had been decided. We started the four-hour drive home at dusk, exhausted but excited, prattling on like children who had not yet developed critical faculties.

Neither one of us slept much that night.

\* \* \*

The next day dawned cool but clear. Unfortunately, we were not nearly so lucid.

Carl fretted through breakfast, trying to talk himself out of what he was about to do. I made small talk (my specialty), which nevertheless always seemed to come back to the idea of buying the ranch. Whatever fantasies he had regarding the land, they were no match for my own. I had the whole place redecorated in my mind before my second cup of coffee.

"By the way, what does the house look like," he asked.

"You're kidding, right?"

"I was so busy looking at the land I didn't even notice it," he said, chewing thoughtfully on his toast.

"I'll take care of that part," I said.

The house was really a doublewide trailer, onto which had been grafted a good-sized great room and a large master bedroom, both paneled in knotty cedar. Not exactly a manor house, but it would do.

\* \* \*

Immediately after eating, Carl drove all the way back to Garberville and made an offer.

By the time reason returned, we were the ignorant but happy owners of Twin Creeks Ranch.

We were not to remain ignorant for long.

\* \* \* \* \*

## CHAPTER 2

# Martha Stewart Doesn't Live Here

*"Fashion fades, only style remains the same."*
— Coco Chanel

*"Art is the imagination at play in the field of time.*
*Let yourself play."*
—Julia Cameron, *The Artist's Way*

PERHAPS WE SHOULD HAVE REALIZED SHORTLY AFTER we calmed down that quaint barns often look that way because they are in an advanced state of decay, and flower-filled meadows appear quite different in the searing heat of summer than they do in the damp greenness of spring.

We haunted thrift stores, garage sales, and flea markets in our effort to transform the empty seven-room ranch house into a comfortable retreat.

We had no budget, so the effect was more Early Student Apartment than Stately Country Home—brick-and-board bookcases, mattresses on the floor and posters thumbtacked to the walls.

This accomplished, we began a little remodeling.

The "two-story guest house" turned out to be an unfinished shell with no bathing or cooking facilities but lots of "potential." Wisely, we decided to worry about that later.

The house was lovely but dark as a mouse hole. Most of the walls were covered with walnut brown wood paneling that absorbed far too much light.

Concluding that camouflage would be our best hope for transforming mouse hole into household, we set to work diligently with white paint and bright wallpaper.

As soon as we applied the kitchen wallpaper, it began to peel off the walls. We had to strip and re-prime each panel before the darned stuff would stick.

We managed to accomplish most of the transformation in one frenzied month of weekend labor, and I think we both showed admirable restraint throughout the entire, frustrating process. When we finished, we were still speaking, but barely.

It also turned out to be a much longer project than we had planned.

\* \* \*

### LETTER FROM ZENIA
### (Excerpts From a Country Diary)

February 12th, 1991, Tuesday

I am back where I belong, having left half of my double life behind rather hastily.

Carl's 55th birthday was on the 9th, and as I am the president of my leads club back in Rohnert Park, I had to attend the Leads Tea today, a once-a-year big deal for all of us concerned. I left at 3:10 P.M., and got to Twin Creeks at 7:10, racing to be with Carl on his special day.

It was a very romantic evening, with candles and flowers and bubbly, even if he did have to make his own birthday dinner.

The leads ladies insisted I take my flower arrangement from the tea with me as a gift to Carl; so touching. I felt like a girl escaping to a lover's tryst.

The ladies of the leads club were understanding, and let me out of my presidential duties early. I bought the champagne on the way up, and Carl had a glass or two of the late-harvest Chateau de Baum wine, from their Symphony grape—delicious!

We talked about being grateful, and about our seven wonderful years together. What a miracle that has been—and is.

February 17th, Sunday Night

Spent the day getting ready to paint the big bathroom while Carl dormant-sprayed the fruit trees.

After a nice walk together so he could show me a new fern he had found, and after we had found a skunk's skull in the woods to add to our bone collection, we talked for a long time.

Monday, February 18th

Today I painted. I did two coats of my specially mixed paint. It's a near miss, but it will be okay.

Tomorrow, I do the cabinets. Marvin and Nancy visited for about half an hour yesterday, to see the place and visit the cows—more about how we got the cows later.

I have to feed Big Mama and Hortense today. Called and called the others but they didn't come. Maybe tomorrow.

I've decided to let go of worrying about it. Marvin says they never forget how to get home. I hope he's right.

* * *

Often we would sit on the deck after a hard day's work, staring into the dark, too numb to enjoy the dazzling display of stars or notice the velvet quiet that enveloped the place.

Some days though, we quit work early and took long rambling walks, discovering frog ponds, bee trees, waterfalls and springs. There were giant boulders covered with furry mosses and many-colored lichens, swamps and secret places, blackberry thickets as tangled as seaweed beds at low tide.

An awful lot of blackberry thickets.

Blackberries, as any gardener knows, are a mixed blessing. The fruit is wonderful to eat on the spot, sun-warmed and oozing sticky purple juice. It is delicious in pies, jams, jellies and cobblers.

But the wild vines have a habit of growing where they are not wanted: in flowerbeds, vegetable gardens and paths.

Ours hung like prickly snakes from every tree; grew out of dead stumps, and layered themselves along the ground.

The poison oak did likewise, often planting itself in the midst of a tangled blackberry patch, making it nearly impossible to pick berries without the sap of the poisonous plant creeping into your freshly scratched wounds, so that one would have a serious case of poison oak as well.

Carl began carrying his machete and a golf-club weed whacker as we took our walks.

\* \* \*

The week we moved in, I planted climbing red Don Juan roses by the big gate at the entrance to the ranch. I could see the photographic spread in *Country Life* magazine already—the lush red and green of the roses against the whitewashed rustic fence: the registered Twin Creeks cattle brand sign on the gate, that I had painted red and green to match the roses; the wildflower meadow and forest beyond; the craggy purple mountains in the distance—it was perfect.

That night I awoke in a sweat. I had forgotten how much deer are said to love roses, or "deer candy," as the locals call them.

"Honey, wake up!" I said, shaking the unsuspecting one. "Wake up!"

"Whaa?" he said. "What is it? Do you hear something?"

"No, of course not," I said. "We're all alone here. I just realized that if we don't put wire cages around our roses tomorrow, the deer will eat them! Isn't that awful?"

"Don't ever do that to me again," said Carl, pulling a pillow over his ears. "Ever."

\* \* \*

At last we were settled, and we couldn't wait to share our beautiful sanctuary.

Soon, we were inviting numerous friends and relatives to visit.

Most actually made the trip at least once, but lost heart on discovering it took four or five hours from where they were to where *we* were.

Still, we did a lot of entertaining that first year. We cooked some great meals, got lots of volunteer help with our large vegetable garden, and even had a little time to enjoy the place. "We won't change it much," we promised each other. "It's beautiful just the way it is."

\* \* \*

Before you could say, "What were we thinking?" we had planted redbud and pine, dogwood, more roses, perennials for cutting, a few more fruit trees, lots of vegetables, and a half-acre herb garden.

To support these we needed a water system—in fact, several. Soon we were buying timers and valves and hoses and of course, more plants.

As anyone with half the critical thinking ability the Good Lord gave him (or her), could see, things were already wildly out of control, but we were too naïve to know it yet.

\* \* \*

March 31st, 1991, Easter Sunday

Got back on Friday afternoon after a short trip to town, arriving just before dusk. Cats here to greet us, and Baby and Hortense and Hamburger, the steer—no sign of the other three cows. "The girls are being naughty again," as Carl says.

The two small ones didn't show up till Saturday morning. We got up early and meditated.

Carl fasting, so I ate alone.

He spent time bustling about in the kitchen, getting the meal ready for our Easter Sunday feast. I was completely out of it with a case of hives—no help at all.

I did manage to walk down to the waterfalls on Saturday. Burgess Creek was a raging torrent because of recent heavy rains.

And how we needed them, as we were headed for the worst drought on record.

Still not many wildflowers blooming, but the little purple-and-white ones and the Baby Blue-Eyes have made the upper pasture a multi-colored patchwork carpet on a green background.

Spotted some wild columbine, and what I hope is a wild orchid. Wild iris blooming on the way to Zenia, but ours are not in bloom yet. Lilacs and fruit trees all at varying stages of bud. No asparagus yet.

Still no sign of three cows. Carl thinks he sees them on the Chambers ranch, adjoining our pasture. Hope they show up soon.

Carl made an Easter egg for each of us, dyed with onion skins. I set the table with his cut-glass goblets and I picked fresh daffodils from the garden for the centerpiece. We had breakfast in the living room so we could look at the view—champagne and lots of hors d'oeuvres.

Haven't accomplished much.

We are still reading to each other, from Gerald Durrell's, *My Family and Other Animals*. So funny.

\* \* \*

I will plant my garden tomorrow if it doesn't rain. Brought up herbs to start my kitchen garden—also flower seed.

I have finally shaken off a persistent flu bug and feel better.

A new beginning.

Forgiveness and resurrection. These are the themes for this lovely day.

And oh yes—we saw the Easter Bunny!

\* \* \* \* \*

CHAPTER 3

# Water

*"And Noah he often said to his wife when he sat down
to dine, I don't care where the water goes if it doesn't
get into the wine."*
—G.K. Chesterton

WITH OUR SECOND SEASON AS COUNTRY SQUIRES
came the realization that all that rustic charm carried with it
a hefty price tag. Those lovely patterns on our knotty cedar
ceiling were in fact water stains, which became more artistic
and pronounced with every passing rainstorm. The roof
leaked like a sodden sponge.

The beautiful weathered barn wood was actually
closer to rotten. The plumbing and water systems kept
breaking down whenever we happened to have weekend
guests, and the elderly water heater rumbled to itself like an
angry volcano threatening eruption.

At first we counted ourselves fortunate to have water
in such abundance. Despite a protracted drought, our pond
still beckoned liquidly in the summer heat, and the two
streams still flowed—perhaps not quite as loudly as they had
the previous season.

We had a number of naturally gushing or oozing
springs on the place in addition to the one from which we
drank and washed and watered, so we felt safe.

In late August, I entertained two members of my sub-
urban women's group for a three-day weekend, and for a
while, we were having a wonderful time hiking, gossiping,

drinking wine and cooking fabulous meals with my fresh garden produce.

"Oh, my God, just *look* at all this basil!" said Janice, the chef du jour, as she gathered armfuls of the pungent green. "I'm going to make fresh pesto for dinner tonight."

It was after that memorable meal, while Pat (the prep cook), and I (the scullery maid), were washing dishes, that the difficulties began.

We were halfway through a stack of pesto-smeared white china when the water in the kitchen tap suddenly gurgled to a stop.

"What happened to the water?" asked Pat.

"Not to worry," I said. "One of the cows probably stepped on a pipe. I'll have it fixed in no time. Have some more wine."

Off I trudged with my backpack full of pipe clamps and screwdrivers.

It was a brisk cross-country hike to the holding tank as I traced the water line back to its source on the mountain behind the house. I could see nothing wrong so far.

\* \* \*

By the time I reached the tank I was winded and hot.

I could tell without looking inside that it was empty. I gave it a hopeful thump anyway. Muffled, hollow sounds.

I started for the spring, sweating and grumbling.

"Ninety-five degrees in the shade and me without water," I said, feeling—as usual—victimized by Fate.

Another quarter hour of sweating and grumbling and I was at the spring.

"That's odd," I said—though there was nobody else to hear except a doe that was vainly attempting to blend in with the underbrush. "No water coming out of the overflow pipe."

Me Embracing 2500-gallon Water Tank

I opened the lid of the spring box. That too was empty.

"Now what?" I asked the doe.

No answer forthcoming, I trudged back to the ranch house, where I heard giggling from the back garden. Obviously I was the only one moved by the gravity of the situation.

The ladies had discovered our backup system—nonpotable water we kept for watering the flowers—and had stripped naked on the lawn and were happily squirting each other with the garden hose.

Fine. I was hot and tired, and they were frolicking.

"Get dressed," I said. "We're going to the store."

I loaded the water nymphs into the car with half a dozen plastic water jugs and we headed out the gate.

The Kettenpom General Store was nearly 10 miles away down a dirt road, which was either very dusty or excessively muddy, depending on the season.

Still, a trip to Kettenpom was more convenient than a trip to Garberville or Mad River over treacherous mountain roads; that is, when the store was actually open.

During the years we owned the ranch, the KGS had seen three or four different owners, all of whom had tried valiantly to serve the small, diverse community. But times were hard in the mountains, and too much credit was often extended, leaving proprietors without the necessary positive cash flow to stay in business.

At the time of the water emergency, however, it was both open and well stocked. The ladies purchased another bottle of wine for emergency consumption while I explained our sad story to Steve, the latest in a series of saintly proprietors, who graciously offered his garden hose.

We filled our jugs and returned home to drink wine and ponder the problem of the disappearing water.

Finally, I caved in and called my husband, but he had no immediate solution either—in fact, I thought he sounded a little hysterical. Perhaps if he had just had a few glasses of wine before receiving the news it would have seemed less catastrophic.

There was nothing for it, he said, but rationing.

I hung up the phone and passed the wine bottle. Then, I decreed that each of us would have two gallons of bottled water apiece, for all necessities save one. The toilet was to be flushed once a day with water from the garden hose.

The ladies groaned, but I told them to think of it as camping, but with softer beds.

\* \* \*

Our first winter was unusually cold, and we arrived once for a relaxing week to find the water pipes had frozen solid and burst in the one place we couldn't reach—under the cement garage floor.

Carl crawled under the house in subfreezing weather, cursing mightily as well he should, and tried to bypass the pipe. A couple of hours of that, and I had to unfreeze *him*.

Of course without water, we couldn't wash the dishes. Or take a shower. Or flush the toilets.

The hose through the bathroom window trick worked okay, but turned the bathroom into something resembling a frozen food locker.

Meanwhile, we drank boiled snow.

\* \* \*

Cows were always a menace to the water system, such as it was. They perpetually stepped on the exposed plastic pipes and crushed them, creating spectacular geysers followed by equally spectacular drops in water pressure at the house.

I remember fondly the old oak stock tank (in a former life, the bottom of a very large wine barrel), which stood in the corral near the hay barn under a spreading black oak tree. It was perfect, that scene, and I hoped one day to paint it, just as it was.

But one morning, gazing out at the view, I sensed something was not quite right.

I groped for my glasses and the binoculars, and peered down toward the barn.

The old oak stock tank was missing!

Alarmed, I hauled on my barn boots and hiked down to the corral to see what was what.

The stock tank lay on the ground, opened like a daisy that had untimely bloomed and as quickly died.

The cows had trampled the water line leading to the tank. The water had then evaporated and the tank had dried out.

When this happened, the boards shrank, the staves dropped to the bottom and the whole thing collapsed outward.

Carl drove up that evening, and we repaired the broken pipe. But we still had no water tank for the cows. So, we did what all good ranchers do in a pinch—we improvised.

We found an old claw foot bathtub left behind by former owners in a neat but odd collection of items we referred to as, "the bone pile," and hooking the tub to the rear bumper of my trusty Diahatsu Rocky 4 x 4, hauled it down to the corral and heaved it into place.

After an initial period of prodigious but disdainful sniffing, the cows accepted their new watering hole, and we drove back to the house to have a tall, cool one of our own.

\* \* \* \* \*

## CHAPTER 4

# Cows

*"Cows are my passion."*
—Charles Dickens: Domby and Son

SPEAKING OF COWS, YOU MAY HAVE GUESSED BY NOW that Twin Creeks was a cattle ranch. If you did you are smarter than we were, because somehow we failed to grasp this fact right away.

In an act of unexpected generosity, Marvin had included cows in the sale of the ranch.

We had been quick to realize the barns would be useful for storing our collection of junk, but slow to realize that cows went with them.

The evening after we had made our offer, Carl said to me, "You know, I think he means to throw some of those cows in along with the property."

To which I replied, "What cows? Oh, you mean the ones we saw wandering around in the meadow? Oh, don't be silly, Honey. What would we do with cows?"

He looked at me. I could tell I had him there. As a city boy born and bred, Carl was not even sure which end of a cow got the alfalfa.

"We're not going to be there all the time," I said shakily. "How many cows are we talking here, anyway?"

"A couple, I guess. It's hard to tell," said Carl, stroking his chin reflectively.

He was uncharacteristically vague. I began to feel uneasy.

"...and Saint Francis had birds!"—Carl

"Well, they are kinda cute," I said. "But what would *we* do with cows?"

\* \* \*

After Carl had come home from signing papers, we celebrated with champagne that night.

In the midst of the bubbly and good feelings, Carl said, "Oh, by the way, the deal *did* include cows."

"My God!" I said. "How many?"

"The whole herd," said Carl. "Isn't that great?"

Great! The herd consisted of three cows, two heifers and one very obstinate steer—but, mind you, the difference between cows, heifers and steers was not even clear to us at that point.

The deal also included two ancient cats that had lived their entire lives on the ranch, and "couldn't be moved."

They were sweet old things who knew how to hunt before they opened their eyes, which turned out to be fortunate, because every attempt on our part to leave food for them when we were gone met with failure.

We purchased elaborate self-feeders (water was never a problem) which raccoons and skunks and possums promptly licked clean. We tried feeding the cats on the roof, but the wild critters found the food there, too.

Beaten, we left the poor pussycats to their own devices. This meant that our field mouse population stayed under control, but the bird community also suffered.

The pussycats stayed fat.

Marvin had wanted to throw in his elderly dog on the deal as well, but we balked at that. *Somebody's* life was definitely getting simpler, but it wasn't ours.

\* \* \*

But back to the cows.

Before I could say, "Gee, they look awfully fat, don't

"Hamburger" the Clown

they?" we were the proud parents of three calves.

That was just the beginning. At their creative peak, we had 26 of the best-fed, best-loved, and least trainable pets on our mountain.

Ours were Polled Herefords, which means they have no horns—a laudable attribute in a creature that tips the scale at well over 1,000 pounds.

We fell in love before we could distinguish one from the other.

Naming them would help us in telling them apart, we reasoned. This was true, but it made the inevitable part-ings more painful than we expected.

There was Buttercup (and her offspring, Lily and

Pansy); Big Mama and Baby; Hamburger and Trouble (self-explanatory); Hortense; Latte her calf, Ghost; Salt and Pepper; Spats; Mocha (Latte's daughter, of course); Curly and Paco and Peaches and Pixie, and many others now gone off to that last great roundup; yet we fondly remember them all.

Our relaxing retreat had quickly transformed itself into our weightiest responsibility.

Soon we were making the eight-hour round trip on a regular basis, feeding animals, watering gardens, pruning fruit trees, repairing farm equipment, fixing the leaky roof, doing general maintenance and trying vainly to keep up a full-time working ranch. We traveled a lot, so this was taxing, and we now had *two* houses and *two* gardens to fret over.

Between trips, we worried.

\* \* \*

We discovered to our relief that cows are relatively self-sufficient. With grass plentiful and water everywhere, they only require the occasional salt lick to manage.

But in winter, the grass grows slowly—if at all, ponds freeze, and heavy snow blankets the pastures. Especially in winter, cows need hay, and preferably alfalfa hay, the most expensive, most nutritious, kind.

Our cows regarded alfalfa with the ardor we reserved for a good filet mignon (though this would be indelicate to mention), and they consumed staggering amounts of the stuff in a single day.

When we could not get to the ranch because of snow, we were reduced to calling our country neighbors and begging them to throw a little hay to our orphans. We began to owe favors to everyone on the mountain.

\* \* \* \* \*

CHAPTER 5

# What Cowboys Do

*"Foot in the stirrup and hand on the horn,*
*Best damn cowboy ever was born."*
—The Old Chisholm Trail: Anonymous

CONSIDERING WE'D NEVER BEEN TO VETERINARY school, I think we did pretty well with our new and unwieldy pets.

We started reading books on cattle raising, exchanging the cocktail party circuit (never a favorite of Carl's) for potluck suppers at the Community Hall, where we cornered unsuspecting cattle ranchers with earnest questions on animal husbandry and the latest remedies for prolapsed uterus, mastitis and scour. We inoculated, branded, castrated, tagged and midwifed.

The trick with such operations, of course, is to get close enough to the cow to perform them. This means luring each cow in turn into a metal contraption called "the squeezer."

It was my job to do the luring—Carl's to do the squeezing.

Once led, pulled, pushed and cajoled into the squeezer, each cow would let us know how strongly she disapproved of this treatment by bellowing and repeatedly heaving her considerable tonnage against its sides attempting to kick someone—anyone—in the process.

A distraction was necessary in order to keep them still enough for Carl to do his work.

That was also my job.

Distraction involved a variety of techniques, from

Carl with Branding Iron

soothing tones to solicitous patting, and if all else failed, holding alfalfa just out of reach of gnashing teeth. There was probably an easier way, but we didn't know what it was.

It always pained poor Carl to cause the cows any amount of distress.

That worried me too, of course, but I secretly feared that at some point they might try to get even.

We brooded for days before each roundup; both dreading what was to come.

\* \* \*

One summer, my daughter, Jamie, brought her Jewish New York boyfriend—a stand-up comic—for a taste of the country life. I don't think he'd ever seen a cow before, and he thought they were funny.

He assumed that they would think he was, too.

Not long after he arrived, he changed into shorts and

canvas shoes and announced he was going down to the barn to have a word with the cows.

They were apparently a tough audience.

He hadn't been gone long when I glanced down toward the barn to see them starting to bunch up around him, backing him into the manger.

I yelled, "Don't let them get you into a corner," but it was too late. An agonized scream issued from the barn.

Jamie and Carl rushed down to the barn to help the injured man limp back to the house...

A foot that has been stepped upon by an animal weighing over half a ton is not a pretty sight. Fortunately the ground in the manger was soft and sort of squishy—had it been hard, his foot would surely have been broken.

Still, the boyfriend had to be driven down the mountain to the hospital for an x-ray, and spent the remainder of his visit languishing on the sofa with his foot elevated and covered with ice packs, consuming therapeutic doses of beer.

\* \* \*

Me Distracting with Hay

Ask anyone who has ever lived near a cattle ranch, and they will tell you angrily about their ravaged corn patch or ruined flower garden, their once-lovely apple orchard or trampled lawn.

On one of my routine visits to our quickly burgeoning herd, I called the cows and counted noses only to find six of them missing.

I knew they were somewhere near the lake (about half a mile from the hay barn), because I could hear them mooing, but no matter how much I called they refused to come.

It was a very hot day and I was not in the mood for a hike. Grumbling, I headed toward the sound, first stopping at the hay barn to stuff a few armfuls of alfalfa flakes in my backpack to entice the cows to come back where they belonged.

Crossing the bridge over Burgess Creek, I started up the steep hill behind the lake, arriving in full sweat at the property line.

Scanning the neighbors' pasture, I still couldn't see any cows. I leaned over the gate and called.

I listened carefully. No moos.

Easing the gate open I called again, and was greeted not with moos, but with a cheerful burst of donkey braying.

Three donkeys came pounding up out of the neighbor's swamp, braying wildly and heading straight for me and my alfalfa.

My errant cows were right behind them, but it was obvious that the donkeys would get to me first. Quickly, I slammed the gate shut.

I would have to try another strategy.

I decided to wade in among them and push them toward the opening in the gate, keeping myself between them and the donkeys and scattering a trail of alfalfa crumbs. This netted me three of the younger ones, but the older cows were more sophisticated. They knew enough to hold out for bigger treats.

There was nothing for it but to hike back downhill, past the lake, across the bridge, and up to the hay barn for more alfalfa. It was getting hotter by the minute. I wrestled two flakes (the 3-4" thick compacted end of a bale—pretty damned heavy) of nice, green alfalfa to the floor of the hay barn and scooped them up in my bare arms. It scratched and tickled.

I cursed the wayward cows as I toiled back uphill to the neighbor's makeshift gate.

I didn't see any donkeys as I approached, but I didn't see any cows either.

Placing a flake of alfalfa on either side of the gate, I opened it wide and called softly to the cows.

Again the donkeys boiled up out of the swamp, braying gleefully. This time I fended them off long enough for our cows to get the idea and bolt through the opening, leaving the disappointed donkeys in the dust.

\* \* \*

Of course, the gate swings both ways.

It was not unusual on doing the nose count to find I had a few extra noses. One morning I noticed we had two more than we were supposed to have: a black, whiteface cow and her young calf.

They had attached themselves to one of our cows who had a calf of her own, and the two calves, both only days old, were frisking happily together in the spring grass as the mothers, heads together, appeared to be swapping calf-rearing stories or comparing notes on the hardships of pregnancy in winter, each cow glancing fondly now and then at her lively offspring.

On another occasion, Carl and I took a leisurely stroll down to the lake for a little birdwatching, only to find we had more than birds to deal with.

Casually grazing their way around the lake as if they belonged were 20 alien cows.

"Holy shit, they're eating all my grass!" said Carl.

We raced back to the house and called the rancher whose brand they bore. He promised he would send someone to fetch them the next day.

The someone turned out to be one of our charming local cowboys, a pleasingly lithe and muscular blue-eyed wonder known as Jimmy Bascomb.

"Yup," he said on the phone, "we'll get on over there about 8 o'clock tomorrow."

Sure enough, he was there precisely at 8 with his loyal sidekick, Jeannie—as sweet and feminine as Jimmy was taut and manly.

"Nice rig yuh got there, ma'am," Jimmy drawled as he looked over my small Japanese SUV—a lightweight compared to the big American pickups favored by the local cowboys.

"Thanks," I said, chafing a bit.

When someone calls me, "ma'am," it makes me feel a little strange—also more than a little old.

"You can just call me Mary Lynn," I said, offering coffee.

The four of us then sat down for some serious visiting.

\* \* \*

Country living is neighbor-intensive, and visiting is a necessary part of life, which can put a very large dent in one's schedule.

I mention this because I discovered after a few months on the ranch that the concept of schedule was not something with which these folks were familiar.

This was actually a good thing, but I fought it at first. It took me a while to realize that life was much more about experiencing and relating (the New Age crowd calls this "being in the moment"), and less about making lists and trying to make people fit your plans.

\* \* \*

Carl and Jimmy engaged in a lively discussion on the fine points of animal husbandry while Jeannie and I talked about whatever it is that women talk about (mostly men), and by the time we finished the pot of coffee it was time for lunch.

I did a little rummaging in the pantry and came up with enough canned tuna for sandwiches.

After lunch, the two excused themselves and drove off to saddle up their horses and fetch their herding dogs.

Carl and I had never seen anyone herd cattle with dogs, and as our elusive schedule was shot for that day, we decided to hike back to the lake to watch the roundup.

Not long after we arrived we could see the couple in the distance, mounted on their horses and picking their way carefully downhill through oak and scrub, surrounded by a pack of panting canines.

The warm sun bathed us as we made our way toward them. There was no sign of the trespassing cows. Had they gotten smart and gone home?

As Jimmy and Jeannie came into view they began to whistle to the dogs, who took off as a single unit toward a low draw overhung by greasewood and giant blackberry bushes.

Suddenly there was a flurry of dust and barking as the cows, which had been resting peacefully in the shade (digesting our superior grass), came bellowing lustily into the sunlight and headed for the far side of the pasture, with the dogs in enthusiastic pursuit.

Using nothing but hand signals and whistles, Jimmie and Jeannie directed the dogs to encircle the herd and drive it homeward.

It was one of the most impressive examples of teamwork between man and dog I have ever seen, and in less than 10 minutes, it was over.

Carl Doctoring "Hortense"

\* \* \* \* \*

## CHAPTER 6

# The Earth Mother Experience

*"Little Boy Blue, come blow your horn,*
*The sheep's in the meadow, the cow's in the corn."*
—Nursery Rhyme: Anonymous

*"Mary, Mary, quite contrary,*
*How does your garden grow?"*
—Nursery Rhyme: Anonymous

SOMEWHERE IN ALL THE FOOLING WITH INTRACTABLE cows and fiddling with the temperamental water system and fighting with the ubiquitous weeds, I developed a passion for herbs.

Not just a few herbs, but nearly 150 varieties. And not just one aspect of herbs; no, it was not that simple with me. I loved everything about them, and wanted to learn all the herb lore I could absorb.

I guess it was suddenly having all that land that made me snap. That, and the fact that herbs were one of the few things cows didn't eat.

First, it was the vegetable garden.

Oh, I've always had one, but I had the kind that allowed two or three people a limited selection of veggies, plus, of course, all that zucchini I'd have to find homes for throughout the summer.

But this garden was huge; roughly a quarter-acre.

The table grapes and asparagus bed were already established when we moved in, and we looked at all that space the first summer and were sad to see it full of weeds, but we had no time for a garden. You can't be a part-time farmer, after all.

We clung to that conviction all fall and right up until winter when the seed catalogs started to arrive. Then, like die-hard gardeners everywhere, we started drafting elaborate garden plans and arguing over which varieties of sweet corn to order. Sweet corn varieties. Plural.

This would be the mother of all vegetable gardens! In addition to corn, we would have potatoes, onions, peppers, leeks, melons, two kinds of radishes (both French and Japanese), pole beans and bush beans, peas and their sugary snap pea cousins, spinach, chard, beets, carrots, four kinds of lettuce, lemon and pickling cucumbers, five kinds of tomatoes and three kinds of squash.

And herbs.

I wanted big beds of cilantro, dill, basil and borage.

Next, I snuck chives, parsley, thyme, marjoram, salad burnet and sage into the flower garden. From that point on, we really lost control.

"Wait a minute!" I said. "How are we going to keep this stuff watered when we're gone?"

"No problem," said Carl. "We'll just hook up the water line to a timer and run plastic pipe down among the beds."

Carl in a Rare Moment of Rest Among the Veggies

Me with Fresh Twin Creeks Garden Veggies

It worked, too, so long as there was water available and the cows didn't step on the pipe and the backup batteries worked when there was a power failure (which happened often in the mountains).

That gave us yet another thing to worry about, and we tried to show up at the garden regularly to weed and tend and fertilize and harvest, but the weeds seemed to grow overnight.

They had a lot of help the first year.

In search of the quickest solution to our water problem, we put in sprinklers. That meant the whole sprinkled area got liberal amounts of water.

Our vegetable garden was upwind from the cow pasture, so every windblown weed seed was soon comfortably at home there.

\* \* \*

The following year, we converted to a drip system and mulched heavily between the beds. It looked much tidier at first, but it didn't slow the weeds down much.

The garden looked gorgeous for the first two weeks after planting, but by harvest time the weeds were waist-high and the harvesting process often resembled an Easter egg hunt, complete with squeals of delight when an actual vegetable was sighted.

And, of course, there was the ongoing battle with deer, foxes, skunks, raccoons, moles and pocket gophers.

In spite of everything, we managed to harvest enough veggies every year to feed a host of hungry people with a lot left over (and that was not even counting the zucchini).

In spite of everything, that is, except the cows.

Our first inkling of trouble came the second winter, when we decided to top-dress our freshly mowed asparagus patch with hay, to preserve moisture and keep down the weeds.

We were busily spreading bales of the stuff right up to the fence line when we heard snuffling noises and looked up to find we were not alone.

The herd was just outside the fence, planning its strategy for a communal attack on the freshly spread hay.

We waved our arms and yelled a lot. Things like, "Shoo!" and "Go 'way!"

The cows were not the slightest bit impressed. We finally had to fence them out of that part of the pasture altogether until the seductive smell had dissipated.

But nothing had prepared us for what we found on our arrival from town one crisp, clean fall day.

I was unpacking the car, and Carl went down to the

barn to put out some hay for the cows. He called them, and soon discovered them milling around near the vegetable garden, not far from the hay barn.

They didn't seem to want to come. This was unusual. Normally, when we'd been gone for a day or so, our bovine friends greeted us whenever they heard us drive in.

Carl walked over and looked a little closer. Something was definitely wrong, but at first he couldn't figure out just what it was. Then he realized the cows were not just near the garden, they were *in* it!

Where the week before there had been rows of corn, beans, tomatoes, and all the rest of our beautiful garden, there was now nothing except a few uprooted garden stakes, twisted tomato cages and stubble.

They had even eaten all the leaves off the grapevines.

Miraculously, they had spared my herbs. There they were, borage and basil and all, standing tall and green and lovely along the fence.

I was so thrilled I went right out and bought some more.

Between planting herbs, baking bread, preserving our harvest, shoveling manure, building fences, developing springs, midwifing calves, cleaning the house and trying to keep up with repairs, we both managed to keep pretty busy on the ranch.

At the same time I was trying to keep my design business going back in Sonoma County, monitoring calls on my answering machine daily and fretting I'd somehow miss the next big design job.

That led to a sort of schizophrenic life—schlepping around the ranch for three or four days in my overalls, doing the Earth Mother thing, then racing back to civilization to wash off the mud, repair my ravaged fingernails, and slip into something silky and glamorous for my urban/suburban role.

I was having an identity crisis. This could not continue.

\* \* \*

May 20th, 1991

Lovely but exhausting two days. Sun shone, birds twittered, cows munched lush grass, flowers bloomed and we labored in the fields in the lovely spring.

Worked in veggie garden all morning yesterday: weeding, fooling with the water system, planting strawberries and tomatoes, peppers, eggplant and Romano bush bean seeds.

For a change of pace, I spent the afternoon weeding and planting in my flower garden. I put in: coreopsis, day lilies, dwarf iris, poppies, stargazer and calla lilies, snapdragons and hollyhocks. Spent the next four hours on my knees, weeding.

Dear God! It was the Earth Mother experience times 10! Sutter's Folly! Archibald's Agony!

Yesterday I found my sitting-place, in the arms of a huge buckeye tree above my flower garden. Shared it with Sir Stubbs, an itinerant tabby who showed up for dinner last night.

NOTE: Had I known then that I was never to have the time to sit in my sitting place again, I might have quit working in protest, but I plunged ahead, all cylinders firing.

\* \* \*

May 21st

Big news:

There is someone in this world who will pay me for growing plants!

My desire to meet new people has led me to join the Women's Farm Bureau, which turns out to be quite representative of the diversity of our little community.

Most of the women are farm or ranch wives, but we also have a sprinkling of artists, musicians, Bay Area weekend fugitives, craftspeople and a healer or two.

Most of the folks in our part of Trinity County have to work a variety of jobs. Unless folks are gainfully retired, they are generally chronically under-employed.

What jobs there are available in our area involve logging, hydroelectric projects, working as ranch hands, or in the underground economy. The Realtor who sold us the ranch held down three part-time jobs in addition to his career as an agent: substitute teacher, paramedic ambulance driver and security guard at the dam. Talk about diversity!

\* \* \*

All I remember about that first meeting was hearing that I could join the cooperative and get paid for growing herbs! It seemed too wonderful to be true.

At lunch later, over the scalloped potatoes, baked beans, and the ubiquitous Jell-O salad, I pledged my life to herbs.

\* \* \* \* \*

## CHAPTER 7

# A Country Fourth of July

*"It is my living sentiment, and by the blessing of God
it shall be my dying sentiment—Independence now and
Independence forever."*
—Daniel Webster

NO COUNTRY FOURTH WOULD BE COMPLETE WITHOUT attending the annual Fourth of July celebration at the Kettenpom General Store.

Our celebration always included live music and a white elephant auction, a barbecue (with beef, pork or chickens provided by a local rancher or farmer), plenty of potluck side dishes lovingly created by the local women, and "fried pies" by the 4-H girls.

Coffee was on the house, too, but most folks made repeated trips inside the store for their beverage of choice, whether soft drinks or beer.

I should probably explain here that the KGS, aside from receiving and passing-on phone messages, was also the local laundromat, with one washer and dryer available in an outbuilding on the grounds, a couple of public toilets and even a shower. Since many people back in the woods did not have electricity or phones, this was a real service to the community.

The Kettenpom General Store was the only public building in our sparsely populated western part of the county (nearly 500 square miles and about 250 registered voters) to be open daily; that is, when they could find somebody dedicated enough and wealthy enough to run the place.

The store and the post office naturally became the local gathering places, so you didn't just go for the mail, you went prepared to visit and catch up on the prevailing gossip. You didn't go to the store for milk alone, but to find out who was ill, who had recently sold his steers and for how much, who was divorcing whom and why, and who was moving out or moving back, and who bought the old Burgess place.

We didn't have street numbers until 1997, so if we wanted a UPS delivery, we'd have to give the street name, and specify: "You know—the old Scott Place."

The other two public buildings we had were a kindergarten through eighth grade two-room schoolhouse, and the truly ancient Grange Hall, which had been rechristened the Zenia Community Hall.

The latter was our community's holiday gathering spot and became the once-a-month theme dinner site when the Farm Bureau ladies decided we needed some variety in our lives. This was the place you went when you were tired of your own cooking (as we had no restaurant or coffee shop), and just wanted to—of course—visit, but on a rather larger scale.

It was the site of many potluck Thanksgivings (held typically the day before the holiday, so we could spend Thanksgiving at home with our families), potluck Christmases and Easters (ditto), not to mention Mother's and Father's Days.

We women especially looked forward to Mother's Day feasts because the men always threw us a pancake breakfast, and it was amusing to see hulking lumberjack-types in aprons, tripping over each other in the small community kitchen but smiling through it all as they dished up flapjacks and bacon.

The Fourth, on the other hand, always took place in a pine grove adjacent to the store and involved everyone— the men tending the barbecue and making beer runs into the store or playing horseshoes and riding herd on the kids; the women laying out the potluck dishes on long tables in the shade of the huge trees on brightly colored tablecloths, and making sure the old folks got fed; the 4-H girls turning out their fried pies as fast as they could in electric frying pans with very long extension cords.

We saluted the flag. The band played old bluegrass tunes while the food was consumed, then cleared the bandstand so the auction could proceed.

\* \* \*

July 4th, 1991:

An incredibly hot 3rd—110° in the shade. We went up to Kettenpom early to help set up.

Carl officiated, making announcements and conducting the auction, which raised $400 for repairs on the Zenia Community Hall—a building which needs a lot more that $400 worth of repairs, and people with time to do them.

We had the salute to the flag; then a chicken barbecue and potluck, the auction, kids' games, horseshoes, the bluegrass band and the obligatory cribbage tournament (the highlight of the day's festivities for many old-timers).

Lots of interesting people, from the Black therapist from Baltimore to the Jew from Los Angeles (these passed for exotica on our mountain), the three Deadheads in tie-dyed t-shirts, and the bearded mountain men who were not often seen in public (these didn't).

Some local arts and crafts (candles, crocheted toilet paper cozies, homemade jams and jellies, baby quilts, etcetera) and a charcoal drawing by Will Barnett—an artist of international reputation, and our community's only claim to world renown.

We bid on Will's charcoal drawing, but ended up with a king-size mattress and springs which Carl insisted I bid on sight unseen (as the new owners would be required to collect it for themselves).

Anyway, nobody else was bidding.

At the time, we figured that a new bed couldn't be any worse than the old foam mattress on the floor of our bedroom, so we took a chance.

The chance we took turned out to lead to yet another adventure in the mountains.

\* \* \* \* \*

## CHAPTER 8

# The Great Mattress Caper

*"Ask, and it shall be given you; seek, and ye shall find; knock,*
*and it shall be opened unto you."*
—Matthew, 7.7

NOW WE WERE THE PROUD OWNERS OF A KING-SIZE mattress and springs in unknown condition.

The next hurdle was to find the home of the folks who were donating the set and schlep it home.

Simple, no?

As it turned out, no.

The Coopers had no telephone, so we were on our own when it came to finding their house (they had left sketchy directions), and we were totally unfamiliar with their part of Lake Mountain, which lay south of the Kettenpom Store—but not to worry.

Gorman Smith (a Los Angeles transplant) was sure he knew how to find the spot—after all, he'd been there once before. Gorman would give us directions if he could ride with us to the Cooper's house, which was some 20 miles down the Covelo road, and thence into the woods where many of the pot growers were rumored to dwell—in other words, no man's land.

As it was, we had to ask directions twice, from heavily bearded men who grabbed their handily stowed rifles and let their dogs bay around us as they eyed us suspiciously.

By the time we got to the Cooper's place, we were all three somewhat tense from those encounters, not to mention that the low-slung, sporty Toyota Supra we were driving had bottomed-out several times on the primitive dirt roads that led to the Cooper's remote cabin.

We honked and hollered, but nobody came to the door to greet us.

"I guess they're not home," said Gorman. "They said to just go on in and take the stuff if they were gone."

"Really?" asked Carl, dubious. (If you did that in his native New York, you'd soon find yourself in police custody—or dead.)

Sure enough, the Cooper's house was unlocked, so we wandered in after a few tentative "hellos" at the front door, climbed the stairs to the second-floor bedroom loft, and helped ourselves to the mattress and springs, hoping fervently that they were indeed the ones we were supposed to take. The thought of ever going back over that rutted dirt road to return anything was more than we could take after wrestling both mattress and springs down the stairs and atop the Supra.

Carl and Gorman and I lashed the king set to the Supra's top as best we could, and began the torturous journey home. Carl drove and Gorman navigated, both peering out beneath the flapping mattress, which either billowed in the wind or snapped down so low over the windshield that driving with his head out the window was a necessity for Carl during the whole trip home.

But first, we had to stop at Gorman's to drop him off, and he insisted on sending for his wife and their camera to record the event.

"Looks just like *Tobacco Road*," he said, delightedly, referring to a Great Depression tale that you had to be over 50 to remember.

We thanked him profusely for his help, as the better part of a day had been spent in the retrieval of our winnings.

Gorman just laughed and said, "Seeing you two in

that car with the bed lashed to the top more than makes up for my time. *Tobacco Road*! Hee, hee!"

* * *

We got home and dragged, pushed and pulled our new mattress set into the house and up the stairs, put our one old king-size sheet on it and flopped down for a well-deserved nap.

We had just dropped off when we heard voices outside our window, calling our names.

"Hellooo the house," they said.

"What the hell?" said Carl, annoyed at being awakened, and as always, alert for trouble.

He threw on his clothes and went over to the window. "By God," he said, "there's a strange man and woman outside. How the hell did they get through the locked gate?"

He opened the window.

"Hey," said the strange man. "We're the Coopers. You know, the ones you bought the bed set from."

"Oh, shit," said Carl. "I knew we shouldn't have taken that stuff when you weren't there. We took the wrong things, didn't we? And by the way, how did you get through the gate?"

"No, no," said the strange man. "That was the one, but you forgot the sheets and pillowcases...

"And the gate was locked, so we just hopped the fence and walked down."

"Come in, come in," said Carl. "We'll be right down."

Turning to me, he said, "Get your clothes on so we can meet the folks, and offer them some tea or something. I can't believe they came all this way to do us a favor!"

But they had done just that.

We gave them some tea and visited. That shot the

rest of the day, but even though our "new" bedding was somewhat more used than we had hoped, it had cost us only $20 and some time, and we'd had a memorable adventure and made some new friends in the process.

After Carl got over his annoyance about the fact that our battlements could be so easily breached, we both slept well that night.

\* \* \* \* \*

CHAPTER 9

# Good Neighbors

July 5th, 1991, Friday

Home from the Independence Day festivities in the late afternoon yesterday, tired but happy. Had salad from the garden.

Today we had to pick up what we bought. That sounded easy, but turned out not to be a simple thing at all.

I sanded director's chairs in order to paint today, and Carl took down the exterior shutters for me. I thought I was going to paint them but they were too far gone. Will have to buy new ones. Flower garden is lovely. Roses and Shasta daisies are blooming. Gladiolus not far behind.

Stir-fry from the garden for dinner: daikon radish tops and scallions, zucchini and small snap peas over brown rice. It was delicious and healthy.

I'm making posters again for Community Fun Night. Always something.

What else? Oh, when I arrived on the 3rd, the cows were up by the small pond. Cisco was standing in the water up to his knees, if cows have knees—so sweet. Not so hot today; only in the '90s.

So nice to have a bed off the floor at last!

July 6th, 1991

Our one-year anniversary at Twin Creeks Ranch.

We celebrated by reading my notes from last year, drinking champagne on the deck and walking to the funny knob that Marvin named "The Teapot Dome" at dusk, each placing a rock under the buckeye tree at the top. Placed two flowers on top of them—I've no idea what they are.

This morning we painted chairs, weeded the flower garden and Carl cut thistles.

Had lunch: a stir-fry of purslane—a weed which is ever-present in our vegetable garden—spinach, chard, onions and brown rice. (We figured if you can't beat 'em, you've gotta eat 'em).

Worked in garden until 6 P.M. More champagne, then dinner and reading to each other, then our walk. Another among many lovely days, but tiring too.

July 7th, 1991

Up at 7. Meditated, ate breakfast and painted two director's chairs for use on the deck and

Ranch House Living Room, 1991.

also for extra seating for the Kettenpom-Zenia Women's Farm Bureau meeting here on the 18th. Then we worked in the garden, weeding and mulching, from 9 to 11:15.

Had lunch, then put the last coat of paint on the bathroom wall.

Took a brief nap, then went for walk around the lake and visited cows, who were napping in a cool grove of trees. They were cool; we weren't.

Back to garden for about one-and-a-half hours. Carl still fighting thistles. Planted one more row of corn.

Dinner, then fooled with wallpaper in kitchen (it's peeling off the wall. Again).

* * *

August 24th, 1991, Saturday

We have definitely lost the weed battle, but pests are not too bad so far, as many pest predators have taken up residence in the garden: the yellow and black orb-weaver, ladybugs, beneficial wasps and garden snakes.

I am hoping for toads, but haven't seen any yet, even though I have provided a lovely toad house near the grapevines. Toad house is now enveloped in weeds as well—perhaps they can't find it for the jungle.

Next year it's black plastic mulch. I never saw weeds cover a hay mulch so fast! So much for the "hassle-free gardening" promised by Ruth Stout in her eponymous book. I curse her name.

A few strawberries. Blackberries starting to get ripe. Deer, quail and bunnies everywhere, but haven't seen or heard wild turkeys this trip.

Our new chaise on the deck affords a great view of early evening's goings-on, but neither of us has spent much time in it yet.

We seem to have forgotten how to just sit.

Yesterday I added fixative and oil to my potpourris—lavender oil to the rose and lavender potpourri, and cinnamon oil to the herbal spice.

No idea how they will turn out, but I am hopeful.

Picked more bay leaves to dry and packaged up some dried lemon balm for tea. Harvested dill weed to dry, and will probably start on coriander seed-drying today. Some of the dill heads are ready now, but we'll be gone for about a week, so expect the rest will have to wait till Labor Day weekend.

Maybe then we can also figure out how to rest from our labors, but I fear the opportunity is long past.

\* \* \*

November 18th, 1991 Wednesday

Our dog, Snazzy, has had many adventures here at Twin Creeks Ranch. There is so much to do: chase rabbits, dig for gophers, eat cow dung, wade in the streams, herd cattle, swim in the lake, have close encounters with skunks, and, of course, chase deer.

She had to be encouraged to do the latter. Carl has spent many mornings out on the deck, nude or nearly so, barking at whole herds that had gathered in our upper pasture during the night and were quietly grazing on our superior grass. (I looked up what they eat, and the encyclopedia said "forbs," whatever they are, but the squire is hard to convince, so he has taken to barking, in—mostly—vain efforts to scare off the deer.)

At first, the dog was only mildly interested in this bizarre human activity, so there was a great deal of Carl's barking to be heard.

Consequently, when Carl went to the post office for mail one morning, and encountered our neighbor, Albert Walpole there, Albert said, "What the hell kind of dog you got over there, Carl?."

"Oh, uh, that was me," said Carl. "I'm giving the dog barking lessons."

"Oh."

Albert didn't know what to make of that.

* * *

Reminds me of the big story—last May, I think.

Snazzy may not have *barked* at deer, but she never had any problem chasing them. One morning I was ranch-sitting alone, as I often did when Carl had to travel.

The dog started whining before I was out of bed, and I figured she just had to pee, so, blearily, I let her out of the glassed-in porch area where she slept, and she bolted for the pasture by the big barns, muttering under her breath. Too late, I saw the four deer feeding there—two does and two small, spotted fawns.

The does leapt the pasture fence, and one of the fawns was clever enough to crawl under it, but that left one helpless fawn trapped in the enclosure.

"Oh, my God," I shouted, and ran out the door, yelling, "Snazzy, no! NO, NO, NO!"

The dog kept on going, with me right after her, dressed in a t-shirt and Victoria's Secret black bikini panties.

Barefoot.

Snazzy kept on, easily beating me to the fawn. By the time I arrived, winded, feet bleeding from the stones and thistles I had traversed in my haste, and yelling all the while, the dog had her mouth around fawn's neck.

They both were startled by my arrival.

Continuing to yell, "Snazzy! NO! Nonononono!" I grabbed the dog's huge head and said, "Drop it!" as loudly and menacingly as I could and, much to my surprise, she surrendered her catch momentarily, but long enough for me to grab the dog by her collar and yank her away.

A brief scuffle ensued, but I finally got the upper hand by sitting on the 100-pound dog until the fawn finally figured it out and slipped under the fence to safety.

When I'd finally limped back to the house, alternately sitting on and dragging the dog the whole distance, the light on my answering machine was blinking.

Who could it be? I wondered. Nobody ever called us there.

There were three messages. Carl had called in a panic, because Angie (Albert's wife) had called him and said, "Something's wrong with Mary Lynn. I heard her yelling at somebody and I called twice, and she didn't answer the phone! Albert's on his way over there now, and I'm headed over with my bolt-cutters, because he yelled and said your gate is padlocked!"

"What's going on?!" said Carl, when I telephoned.

"Well, everything's under control now, but the dog brought down a fawn and I had to rescue it. It's a long story."

"Well, you better call Angie and head off Albert. I'll hang up and you can call me back later. I thought you were being murdered or something!"

I called Angie, then threw on some jeans and my sneakers and left the dog locked in her sleeping porch.

I ran up the road to the gate, to find Albert and

another neighbor, Cindy. Seems she was gonna climb over the gate and check things out, while Albert backed her up with firepower.

They laughed and went home. First I called Angie to thank her.

She said, "My Lord, girl, I was sure you was bein' raped or something. I heard you hollering, and it sounded like you was saying, "Angie! No! Angie! Help!" so I sent Albert over to see if you was okay. Well, anyway, why don'tcha come on over and have a cuppa coffee and some of this fresh coffeecake I just pulled outta the oven. You been through a lot, girl!"

"Thanks," I panted, "but I've gotta call Carl first and fill him in on the story."

"Well, you go on, and then come over to the house and just sit and visit awhile."

I did just that, but it took me some time to calm down and tell my whole silly story.

The next day, I went to the post office, and Lou, the postmistress, said, "I heard you had a little problem with a dog and a deer." She laughed then, no doubt thinking how ridiculous I must have looked. I expect I did.

How grateful I am for good neighbors.

And for how fast news travels through these woods.

\* \* \* \* \*

## CHAPTER 10

# A Cowboy Wedding

*"Come live with me and be my love;"*
—Christopher Marlowe,
*The Passionate Shepherd to His Love*

OF ALL THE EVENTS WE WERE A PART OF DURING OUR time on the mountain—and there were many different kinds: Funerals, memorial services, bingo tournaments, sewing bees, endless meetings about ways to get a garbage transfer station in Kettenpom or Zenia, potlucks and fellowship—the most fun and entertaining was the cowboy wedding.

The invitation featured a lasso twined around the initials of the happy couple.

Jimmy Bascomb, the community's most eligible bachelor, was getting married, and we were invited.

Like most contemporary California wedding invitations, this one struggled to both acknowledge and encompass several shattered and re-formed family units.

I have been to all kinds of weddings and all kinds of receptions, from the hastily conceived to the militarily planned, and most have been boringly predictable.

This one was different.

It was to be a rather formal affair; that meant shiny boots, black jeans, starched shirts and string ties for the men, and dresses for the women—most of them, anyway.

The community of which we were now residents

turned out in strength to celebrate the wedding. (The promise of free food and beer didn't hurt either.)

Jimmy was an itinerant cowboy and part-time rancher who was proudly raising a herd of his own.

His bride, Jeannie, came with two small children (as so many brides nowadays do), and the couple quickly produced a third. The baby sat and burbled happily throughout the entire affair on the lap of her proud grandmother.

\* \* \*

The ceremony was lovely, brief, and touching, at the Lamb Creek Christian Center in Mad River, and marked by a few embarrassed giggles from the bridesmaids, but the bride and groom were obviously enjoying themselves. The minister was an aunt of the groom, and was as nervous as he was.

The bride and her attendants wore satin; all sewn by the ministerial aunt, and the bride's white gown was trimmed with lace made by hand by her grandmother. A family friend took pictures.

The groom wore a black, gray-trimmed tux jacket with a suitably Western cut, and black braid adorning the lapels. The tails of the jacket fell straight and fancy over his stiff, immaculate new black jeans and cowboy boots, and later at the reception, he complemented the outfit with a black felt cowboy hat.

Groomsman were attired in black jeans and short black tux jackets (and of course, cowboy boots), with black string ties and rigidly starched white shirts.

One of the ladies attending the wedding was a walking billboard advertisement. " God loves you, so act like it!" was appliquéd on her vest in white felt. A number of buttons with similar messages emblazoned her chest.

\* \* \*

It was the reception that everyone looked forward to. It was to be at the Community Hall in Mad River, and by the time we arrived, preparations had reached a fever pitch.

Local women had left the ceremony early to lay out the wedding feast (no caterers there). A cowboy band played, and the best man put on his Forty-Niners cap to complete his outfit and, relieved to have the obligatory toast out of the way, energetically whirled his wife around the floor.

The main attraction was the beef Jimmy had grown himself. He wanted it just right for the reception, so he kept disappearing around the back of the hall to check on the coals.

The beer flowed freely, but there was no sign of food for a very long time, and the assembled were getting mildly snockered. Finally the word came down that the coals were ready, and the groom personally supervised the cremation of a great deal of homegrown beefsteak.

The usual potluck dishes showed up—mountains of potato and macaroni salad, plus, of course, the green bean casserole with mushroom soup gravy and French-fried onion rings on top, plenty of Jell-O salads, often featuring mini-marshmallows—but there were also some pretty ambitious offerings: English trifle, made with layers of raspberry gelatin, sponge cake, jam, canned ambrosia and Cool Whip topping—and lots of my personal favorites: homemade brownies and berry pies.

The grandparents on both sides of the families were there, and a large number of babies, so there was never a vacant lap. A great many beery toasts were made, and the mother of the bride read a cowboy poem she had composed herself, to warn the bride of the pitfalls in store as the wife of a cowboy:

> "...Of course, there'll be some drinkin',
> And maybe a fight or two,
> But you know that he'll come home at night,

As cowboys always do."

As the band swung into a determined hoedown, I silently and sincerely wished them happiness. Perhaps the old ways are best after all.

\* \* \* \* \*

## CHAPTER 11

# Letter From Zenia

September 15th, 1992

Found myself in a really silly predicament today.

Left Carl digging up the ground near the holding tank to see where we are leaking water for the household system, and now I find myself still up at our nearest neighbor's house, and not by choice.

I've been here for the last hour, sitting on Cindy's front porch, having narrowly missed being chewed by her dog. Trapped with two comatose, flea-ridden kittens that didn't even wake up when I walked toward the front door, and clouds of wasps and flies circling us.

Grateful there's a place to sit, and happy it's in the shade.

Unfortunately I'm afraid to make a break for my car. I'm inside a gate on the porch, but the dog is between me and escape, and the neighbors are nowhere to be found, though Cindy's car is here. I can never tell when is the best time to visit; they have no phone.

What to do? Carl won't miss me for some time. Should have planned this better. Shouldn't have gotten out of the car. Anxiously awaiting Cindy's return.

Dog sniffing around my little island of safety, growling and snapping fiercely from time to time.

Every time I try to leave, he rushes me, and I retreat again inside the gated enclosure. Carl probably won't miss me for hours.

I have to pee.

September 17th

I'm trying to get caught up with my journal. It occurs to me that I never mentioned the shooting stars. In August, there was a beautiful meteor shower for three nights running, and we sat outside on the deck after dark to watch.

It was truly amazing. It's so clear up here, with no interference from other lights at night, that when we turn off all but the house lights and creep outside in the dark we're able to see much of the night sky plainly with the naked eye, including the Milky Way and the spiral nebula of which it is a member.

On this particular night, our first encounter with the meteor shower in unrestricted space, we could even see colors in the trails left by shooting stars as they streaked across the blackness of the night sky. So many colors: green, red, purple, gold, white and blue.

I got a crick in my neck just watching them, but it was worth it.

There have been other strange night phenomena in the sky above our Zenia home. As we watched often at night to find constellations we could recognize and name, the heavens appeared to shimmer. We could find no logical explanation for this. Are the Northern Lights visible this far south? I think not. But what could it be? The night sky appears to undulate—growing and billowing as though it were ruffled by ebbing and flowing galactic winds.

Whatever it is, we are blessed to see it, for it is truly a wondrous sight.

\* \* \*

October 20th, Tuesday

Yesterday was my birthday. Slept late, and had a late lunch. It was a lovely day—warm but not hot.

White puffy clouds were everywhere.

Found cows by the lake on our evening walk. Fritz, the mother cat, in the orchard; Sparky, her offspring, on the roof. All present and accounted for.

"Sparky" Asleep on Her Feet

Spent the rest of the afternoon repairing leaks in the water line to the pond and waterworks in the orchard, in the lines from the new spring. Water line to flower garden is malfunctioning again. Had a toast on the deck with our sherry, then a wonderful meal of chicken livers and mushrooms in sour cream with capers and onions over rice. Cholesterol, schmolesterol.

It's handy, I find, to have a man who cooks.

Carl gave me a cow card (natch), and a lovely pair of earrings I had admired on a vacation trip to Greece. I was thrilled. Early to bed.

Up this morning at 7 A.M., and heard rain at last! We have been having a terrible drought, and water is scarce. Creeks are down.

Cuddled and snuggled in bed till 8:30, listening to the unfamiliar sound.

This afternoon we cleaned out the gutters around the house. Much fancy ladder-work on both our parts (and lots of praying on mine).

Came back and had popcorn and tea and built a fire.

Tonight we had cranberry beans from our garden.

Still raining. Must be half an inch by now. Our rain gauge is unreliable, as it has sat in the weather too long. I thought they were supposed to withstand the elements, but maybe not here, where we get an occasional snow or hard freeze.

\* \* \*

November 11th, Wednesday

Our little herd is growing! Last night it was 33°, but discovered something momentous had happened during that cold night.

This morning before breakfast—I was messing with the coffee and trying to start a fire in the bitter cold—Carl called the cows and instead of five, seven appeared. Hortense and Buttercup had both had babies. Two little black-and-white bull calves—Hortense's has an all white face and Buttercup's baby has funny black spots around each eye and two in the middle. He's the oldest, I think. He must've been born right after we went to bed last night.

I dreamed of waking up and finding two calves, and here they are. Baby and Big Mama are most certainly pregnant, too.

This will be Baby's first. Must be pretty mystifying stuff for a new mom. (I know it was for me, and I had done a lot of reading before the fact!)

Drinking-water tank is filling up again. Rains got the springs moving, so we have water finally. Walked down to our lake this morning after breakfast and saw about 50 wild ducks on the pond—Buffleheads, I think.

Disconnected the waterworks and moved the pipe out of the stream for the winter. Much clambering around on wet, slippery rocks, with the dog getting in the way the whole time.

Repaired the old sheep barn, then back for lunch and apple-picking in the orchard. Our Granny Smith tree has yielded five bags of apples so far, and many are left.

Planted Egyptian onions, red onions and garlic in veggie garden and dug potatoes. Only five, for all the hills Carl planted. Somebody else beat us to it (gophers, no doubt).

Only Fritz showed up this A.M. for food. No sign of Sparky.

Unloaded gates for a new fence.

* * * * *

## CHAPTER 12

# BabyHolds Out

*"I think I could turn and live with animals, they are so placid
and self-contain'd..."*
—Walt Whitman, *Song of Myself*

SOMEONE MUCH WISER THAN I AM ONCE SAID, "LIFE
is what happens while you are making other plans." JOHN LENNON

It was something I'd never planned to do—live on a
ranch and raise cattle, for heaven's sake! I was an interior
designer, after all. What would ranching do to my hands?
Shoveling manure could give one unsightly calluses, and
would certainly upstage my designer perfume.

But Carl had his dream and, manicured nails or no, I
was swept off my feet by the love of cows.

It was glorious. It was scary. It was fun.

Of course we wanted to be there for holidays, but we
always had family obligations down south that were difficult
to ignore. Thanksgiving 1992, however, Baby had other
plans—plans which we had not anticipated.

It was two weeks before Thanksgiving and Baby, our
first calf, was—we judged—within about a week of being a
mother herself.

In our ignorance we were sure that Baby would
oblige and give birth on cue, in plenty of time to let us get
back to civilization, to our Thanksgiving with Carl's son,
Tim.

We packed enough food in the pantry for the expected lying-in period, but by the beginning of the second week

we were running low on fresh vegetables and milk, and Baby didn't look any more pregnant than when we had first arrived.

We had done all the necessary repairs and pulled all the weeds in the flower garden.

Meanwhile, Baby stayed close to the barn so she wouldn't have to heave her ponderous body too far to get to the hay. She was, after all, eating for two.

By the following Monday, our fresh provisions spent and the canned tuna supply running low, we came to terms with the reality that we needed to make a run to the Kettenpom General Store for iceberg lettuce, oranges and bananas, plus some cheap port and sherry to see us through the cold winter evenings spent around the fire, reading to each other or even talking, things that in suburbia with the ubiquitous television, we often forgot to do.

* * *

We spent a good part of the next morning planning meals for the week to come (just in case), making lists, driving the seven or so rugged miles to the store and, of course, visiting.

By the time we had shopped for canned goods, wine and lettuce, then driven the same dusty, bumpy miles back home, we needed to make lunch.

After an uninteresting salad of iceberg lettuce and unripe tomatoes, Carl went down to the barn to check on Baby.

Alas, she was still gravid as ever, dragging her swollen belly so low it nearly touched the ground. When she lay down, which was difficult in her condition, it was even harder to get back on her feet again.

Having been pregnant myself once upon a time, I sympathized.

"I think she's ready," said Carl with authority when

he returned, breathing hard. "She should have it either today or tomorrow."

"I hope so," I said dubiously. "We've got to leave by Wednesday morning to avoid the holiday traffic."

Wednesday came and went, and still Baby heaved her huge belly up and down around the pasture below the house and stayed close to the barns.

"Well," I said. "At least we can keep an eye on her here. Maybe you'd better call Tim and see if he can come up here for Thanksgiving—and bring a turkey!"

"I guess I'd better," said Carl. "We've waited so long, I hate to go down the mountain now. I know it's going to be soon, and I don't want to leave her alone—it's her first calf. Anything could happen."

"I don't either," I said. "Give Tim a call, but tell him he has to leave by 9 A.M. to get here with the turkey in time to get it in the oven by, say, 1 P.M. You'll have to impress that on him, if he can come. Otherwise we'll be eating at bed-time."

"I will," Carl said. "Is there anything else he needs to bring?"

"Well, let me see. We need cranberry sauce—have him get the kind with the whole berries. There's enough old bread around here to make stuffing, but I'll need celery and a yellow onion and bacon and whipping cream and some kind of green vegetable—frozen peas will do—and sweet potatoes. I've got some canned pumpkin, but have him get me a frozen pie shell, that'll save time, and..."

"Hold it! I'm trying to write this down. Now, cranberry sauce, celery and bread and pumpkin and what?"

"No, not bread. Not pumpkin. I've got those. Oh, and champagne."

"But, of course," said Carl, and somewhat cheered by the thought of champagne, went off to make his call.

Meanwhile, I spent the rest of the evening with my

head stuck alternately in the refrigerator and the pantry, pulling out anything that seemed a likely candidate to add to our Thanksgiving feast.

There wasn't much—the pantry yielded a sprouting onion, ditto several garlic cloves. I found the can of pumpkin and enough frozen bread heels (Carl hoards them for bread crumbs) to make a decent stuffing, and a box of cornbread mix—my Southern heritage (mother) dictated that any stuffing I made have cornbread made with bacon drippings as a key component. The Northern half (my father's side) insisted on boiled, peeled chestnuts, chopped fine. Sometimes they both won.

I figured that asking Tim to find fresh chestnuts would be pushing it, so decided to forgo such an exotic treat that year...

\* \* \* \* \*

## CHAPTER 13

# A Feast of Thanksgiving

*"O Lord, that lends me life,*
*Lend me a heart replete with thankfulness."*
—William Shakespeare

Thanksgiving Day dawned cold and clear. Carl was up early as usual, to walk the dog and feed the livestock.

There was one cow we could always count on to tell us when it was feeding time, and that was Hortense. All the rest of the herd usually deferred to her when any group decisions needed to be made, such as when to head for the lower pasture, when to head for the barn, or when and where to have a delicious afternoon lie-down.

When the cows would head up the hill after a day of grazing and digesting in the lower pastures, she was always in the lead. And now from the bedroom I could hear her, bellowing for breakfast. The rest milled around the barn, letting Hortense handle it.

By the time I rolled out of bed around 7:30, Carl and Snazzy had fed the cows and were off on their walk, rambling over pastures, across streams and through woods.

I was never sure who enjoyed these walks more, the man or the dog. Both always came back panting happily, so it was hard to tell.

In the meantime, I got breakfast. I decided to make waffles, as it was a special day, and I just happened to have enough pancake flour and maple syrup in the pantry. I made my specialty, blueberry pecan waffles, with frozen blueberries from our garden.

I had some ancient bacon that was also frozen, and I thawed that and fried it, hoping to get enough bacon drippings so I could make my cornbread dressing. The bacon was barely edible, but by that time I wasn't too fussy.

Carl came in the front door, shedding his big rubber boots in the entry and letting out much of the heat from my morning fire in the process.

"Close that door!" I said. "How is Baby?"

"Any minute now," Carl said.

I'd heard that one before.

As we were eating breakfast, I said, "Maybe you'd better just give Tim another call, to make sure he's up and ready to go."

"That's probably a good idea," said Carl. "He's probably exhausted, and if I don't call him, he might oversleep."

It was then 9 o'clock, and I was already beginning to worry about the turkey. After all, it was a lot to ask, and even if Tim was up and ready—which I doubted, that would mean it would be 1 o'clock or 1:30 by the time he arrived at the ranch, and 2 or 2:30 by the time the bird was in the oven, assuming it was thoroughly thawed.

At about a half-hour to the pound of stuffed bird, that would mean it would be at least 7 o'clock by the time we ate our Thanksgiving meal.

It was then that I overheard the telephone conversation between Carl and his son.

It went something like this: "Hey guy, did I wake you up?"

Disaster!

After Carl hung up the phone, I started to fret out loud, though this was a bad idea.

"That means 10 or 10:30 by the time he leaves, and that means we don't eat till 9 o'clock at night!"

"Oh, stop fussing. He'll be here."

"Okay," I said.

Around noon, Tim called again.

"I'm on my way," he said. "I'm just leaving Mill Valley now."

"Oh, uh, great!"

As dusk was falling, around 5:30 P.M., we got another call, but not from Tim. It was a stranger calling to say, "Some guy asked me to call you to tell you to pick him up.

"Who?" said Carl, though, of course, he knew.

"Said his name was Tim, I think."

"Why?" asked Carl.

"Said he's having some kind of car trouble. Asked me to call you to pick him up. Got a dog with him."

"Where?" asked Carl.

Oh great. Twenty Questions.

"Said he'd be at the post office in Piercy," said the stranger, and hung up.

This was not heartening news. Piercy was definitely closer, but it was still at least a two-hour round trip by car, and he would have his dog and the turkey and groceries with him (not to mention the champagne, which I was beginning to need badly), and all we had to drive was my little Rocky, which was barely big enough for two people and one dog.

Nothing for it but to pile into the car and eat some chips to stave off hunger.

And that is what we did, leaving the faithful Snazzy behind to guard the house. She was not happy, but at least she had eaten dinner, so I didn't see how she could complain.

\* \* \*

Piercy looked deserted. So did the Piercy post office.

We looked across the dark street and spotted a couple of children playing and dancing in front of a dilapidated trailer parked in a muddy, empty lot.

A man with knotted long hair and a well-worn face answered the door, a plate of hot food in his hand. I willed my mouth not to water.

"Well, some guy was over by the post office for awhile," said the man as he shoveled turkey and dressing into his mouth. The smell of roast turkey, onions and sage was almost more than I could bear. My stomach began to growl audibly.

"What was he doing there?" asked Carl.

"When I saw him he was driving around in circles—backwards."

"For God's sake! Where is he now?" asked Carl, alarmed.

"Dunno," said the man through a mouthful of turkey. "He left."

"When?"

"Oh, lessee. 'Bout an hour ago. I think."

And he disappeared into his nice, warm trailer to finish his Thanksgiving dinner, his two small urchins trailing behind.

"Why the hell didn't he stay put!" Carl said.

"Well maybe he started walking," I offered. "But what the hell could have happened to his car?"

"Oh, he drives those damn wrecks," said Carl, rapidly working himself into exasperation to mask the worry creeping into his voice. "Five vehicles and not one of them reliable."

We drove along the main street of Piercy, looking into every driveway. It was difficult, because there was no sidewalk and there were no other public buildings besides the post office—just one long mile or two of private driveways, most of them dark.

No Tim. He didn't seem to be anywhere. I began to worry, too.

This went on for nearly a half hour, but we were

finally forced to leave Piercy behind and search elsewhere.

Back on Highway 101 at the north end of town, we started scanning the side of the road. Finally, at the Chevron station across from French's Camp, I saw two huddled shapes in the gathering fog.

"Carl, stop! Go back! Turn around!" I said. "It's Tim! It's him and his dog!"

"Where?"

"Just turn around. We passed him!"

There being nobody else on the road at that peak Thanksgiving dinner hour, Carl was able to make a quick u-turn in the middle of the highway.

There they were, leaning against the building and both looking cold and forlorn. We loaded Tim and his dog Rocky into my little Rocky (no relation), along with the Thanksgiving goodies, and, of course, the champagne, and started off for the ranch.

Both Tim and the dog looked a little blue around the edges. The turkey and the champagne were very cold too, so that was one good thing.

"What the hell happened?" said Carl.

"It's a long story," said Tim.

"Well, we've got a long drive, so let's hear it." said Carl.

"Well," said Tim, yawning. "I was heading past Piercy just about dark, and I was coasting down the big hills in neutral to save gas, and when I tried to put the old Chevy back into gear, it slipped into reverse. It would only drive in reverse!"

He then explained how he had backed the elderly car up the 101 freeway and onto the Piercy road, and driven all the way to the post office at the south end of town in reverse, and then back on the freeway to the gas station.

It would have been funny if it hadn't been so dangerous.

"Did that guy call you?" Tim asked.

"He did," said Carl, "but he said you'd be at the post office."

"I was there for about an hour, but I kept falling asleep sitting on the floor and I was afraid I'd miss you."

"Well, we almost didn't see you," I said. "We've been driving around Piercy for the last half-hour peering into driveways. But we found you, and that's what matters."

"Jesus!" muttered Carl. It was the hunger talking.

\* \* \*

"We drove back with the windows cracked, letting in the frigid air. With three people and one dog stuffed into one very small Diahatsu, the normally efficient defroster simply could not keep up, and the windows were so steamed I had to keep clearing the driver's side of the windshield with my cold, wet-mittened hand.

Home at last, we unloaded everything and everybody from the car, preheated the oven, and started our food preparation. While we were doing this, Tim, who was exhausted as usual (only more so after the backwards car ordeal), fell asleep on his bed. The dogs cavorted in the vestibule, and Carl and I cooked.

Carl's brilliant idea was to cut the turkey in half and pre-brown it in a skillet, which we did. Next, we popped the precooked bird halves into a 350° oven and started on the veggies. As the dressing was made the day before and we couldn't stuff the bird, it meant the dressing could bake right next to the turkey.

We started at 7 P.M., and the feast was ready by 9:30—a miracle of Sutter ingenuity.

Unfortunately, we couldn't awaken Tim, so we ate dinner as he slept, straight through to the next morning.

Before our tardy dinner, Carl and I said a prayer of thanksgiving that Tim was with us, and safe.

\* \* \* \* \*

## CHAPTER 14

# Baby Comes Through

*"Life is an end in itself, and the only question as to whether it is worth living is whether you have enough of it."*
—Oliver Wendell Holmes, Jr.

THE NEXT MORNING, TIM WAS REVIVED ENOUGH TO trek down to the pasture to look for Baby, who had not shown up for dinner the night before—something that never happened when we were there.

Carl was understandably concerned when all the other cows showed up for their alfalfa, without Baby.

Carl and Tim came in from the barn about 15 minutes later, huffing and scattering alfalfa crumbs all over the floor.

"We're gonna hike down to the falls and the lower pasture and see if we can find her," Carl said. "We need some water."

I provisioned them and they went off down the path with a backpack of water and a small flake of hay with which to coax Baby back to the barn.

I swept up the mixture of dirt and alfalfa, grumbling. They were gone for what seemed like an hour, and I was beginning to worry about them when they showed up again, still trailing hay and puffing harder now.

"We found her!" Carl said triumphantly. "She's down on this side of Burgess Creek, right by the big waterfall, and she has a calf with her—he's the cutest thing! All black with a white face. Looks like he was just born!"

"Trouble" at "Baby's" Feet

"We're gonna get some more hay and take it to her," Tim said, excited. "She looks really tired."

Tim's enthusiasm for old cars that he planned to restore was second only to his love of animals. As a kid he had snake, fish and iguana pets, and he and his sister Amyre always had dogs and cats.

Now he realized he had the opportunity to enjoy really big pets, and he glowed with enthusiasm.

"I'll bet she's tired," I said. "If I had to give birth to something that big, I'd be tired, too. You're not going to try to move her now, are you? She's so inexperienced, I'm afraid she'd forget what she was supposed to be doing and leave her baby behind."

"No," Tim said. "We're gonna leave her there. We're gonna take her some water, too. She looks thirsty."

"Good. I'll go down and check on her and introduce myself to the calf later on, maybe after lunch, and that way they'll both have a chance to rest.

"But I think that calf needs to nurse right away."

"Right," said Carl. "We'll stay with her and make sure that happens."

By the time they came back, it was nearly lunchtime. I made some tuna salad for the boys. They were so hungry and tired, you'd think they had done all the work of giving birth.

"She was really hungry," Carl said. "We found the afterbirth and threw it way down the hill so the coyotes won't be attracted to the place, I hope.

"You should see where she chose to have her calf—it's a beautiful, sheltered place. I think she knows what she's doing."

"Pansy" Near Sheep Barn

"I hope so," I said, pushing myself back from the table. "Well, I've eaten. I'm going down to see her now. The dishes can wait."

"Take her another flake of hay," Carl said, "and stay this side of the creek and just walk downhill. You'll find her."

I put on my backpack and checked my water supply, then walked down to the barn and filled the pack with alfalfa. The other cows, which were hanging out in the pasture by the barn, eyed me curiously, and those who were no longer dozing or digesting their morning meal heaved themselves up and sauntered over to see if any more hay was forthcoming.

I shooed them away, which wasn't easy because they could smell the hay in my pack, but they finally gave up. I didn't want them tagging along and bothering Baby and her calf.

Hiking down the steep hillside that followed the stream, I soon discovered Baby's den. The noise of the waterfall nearly made me miss her, but over it I could barely hear the soft, lowing noises of our little mother.

She had wanted to be alone, all right. (We later learned that calf birthing is often a spectator sport, and for good reason. A cow and her calf are at their most vulnerable during the birthing process, and the rest of the herd often stands guard to ward off predators such as mountain lions, coyotes and poachers).

The old-timers on the mountain no doubt could have told us all this, as they could have, as it turns out, told us many useful things, but we were too proud to admit how little we knew about cattle and their ways, so we were often busy reinventing wheels, a pattern which we later found to be unnecessary in a place where everyone still depended on everyone else, and friendships were carefully cultivated and held close to the heart.

\* \* \* \* \*

CHAPTER 15

# We Dig In (and Also Out)

*"Better is a neighbor that is near than a brother far off."*
—The Holy Bible, Proverbs—Ecclesiastes

SPEAKING OF NEIGHBORS, LET ME GO BACK A BIT, TO
our first winter at the ranch. We were to learn many lessons
about life and people during our sojourn at Twin Creeks
Ranch, but perhaps none was so telling as our first experience of being snowed in.

Of course, we had been forewarned by the former
owners that if there was any sign of snow at all we were to
drive our vehicle all the way up the hill to the gate (nearly a
quarter-mile), so that when they plowed the road we had
only to dig out the car and sail right through.

Nevertheless, like the tenderfeet we were, we were
immobilized by our first Zenia snowstorm.

A look at my journal from that first winter will perhaps be more indicative of our ignorance than anything—
though I should have been tipped off by those photos of a
snow-covered Zenia we'd seen on our first trip up the mountain. Let me digress a bit:

December 17th, 1990, Monday
      Arrived for pre-Christmas respite. Tired just
thinking about what we must accomplish before
the actual respite begins.

December 18th, Tuesday
      Morning—

Rain. Took walk down hill by big trees and rocks.

Found new ways to get down to creek. Beautiful rocks, ferns, etcetera, glowing in the mist. We wore our yellow slickers and stayed mostly dry.

Sat by the fire and drank hot tea afterwards. Afternoon—

Rained. Harder. At dusk, it started to snow. By midnight we had about a foot. Driveway covered. Toyota stranded next to house. Pipes frozen. No water. Still was beautiful. Cows seemed perplexed, (or maybe that was how we felt). Fed them two bales of alfalfa.

December 19th, Wednesday

Snowed lightly to heavily all day. Walked to pond and back through the stuff. One-and-a-half feet deep by now. Tiring. Feet frozen. Creek rushing. Pond frozen except for 10' circle in middle where forty ducks swam. Saw no deer today. They're hiding. Birds everywhere, especially in barn, twittering, walking around cows' feet, eating alfalfa seeds.

December 20th, Thursday

Still no water. Pipes froze every night. One burst under house. Carl capped it. Got thoroughly wet in the process, froze hands and feet.

Sunny, but not above 20°. Birds tame, letting us approach close to where they browse the frozen ground for seeds. It's like, "We're all in this together." A few bunnies. Loads of quail— maybe 10 in a covey—working the thawed patches under the trees. Temperature at four degrees overnight. Moved bed into living room and closed off bedrooms. Got up five times at night to add wood to fire, then had to read to get back to sleep. Reading *"House of Spirits,"* by Isabel Allende. Depressing.

December 21st

Still no water. Melting snow to drink. Carl started digging out the driveway so we can make our escape. The main road is plowed, but there's no way to get up the frozen, snow-covered driveway in the Toyota.

So we dig. Carl working on his poetry, a Christmas gift to his kids, when not fooling with water system or digging. Our survival effort takes most of our energy. I melt snow, haul firewood, supply tools, and act as runner and general gofer. Tried to write my letter to features editor of *The Press Democrat* about a column, but mostly we spend a lot of time staring out the windows at the snow and thawing our wet, cold feet.

Saturday December 22nd

More digging. More sunshine. No water.

Sunday December 23rd

Dug for six hours today. Finally we're within forty feet of the gate. So is car! We revved up the engine and slipped and slid up the driveway, hoping to get within sight of the gate before it snowed again.

Heard coyotes howling last night. They got something. Hope it wasn't one of the cats, which are mostly keeping warm in the attic, where they continually fight off marauding raccoons.

Ranch House in 2' of Snow

Monday December 24th:

Rose early to finish the digging. I have a headache—the kind that hurts like holy hell when you bend over—so Carl does the last 40' alone. I get things ready to pack up the quarter-mile to the car and close the house.

Carl made a sled out of an old piece of plywood with a rope to pull it, and we dragged our Christmas gifts and traveling clothes up the driveway to the car.

Relieved to learn we have plenty of bales of alfalfa for critters. Crows showed up today. Also deer. Sunny and warm. Snow beginning to melt. We made our escape, down icy roads to Garberville and back to Healdsburg for Christmas. If we hadn't promised family we'd show, we might have stayed happily at the ranch, but must put in an appearance.

Thursday December 27th

Returned to Twin Creeks Ranch. This time with Sylvester (our fluffy black-and-white town cat). He doesn't travel well unless he's in my lap, then his mournful cry is replaced by contented purring. Didn't get here till 10 P.M. Tired. Stopped at McDonald's for dinner. Sylvester freaked by two Malamutes in parking lot. Wouldn't drink water we gave him. No major confrontations after that.

Sparky and Fritz were less than overjoyed to see him. Went to bed at 11, exhausted.

Moby Dick, our one-ton GMC 4 x 4 did fine in snow, but it's always nerve-wracking traveling in that big truck.

Friday, December 28th

Walked down to Big Falls to see the snow and ice. Lovely.

Lake water is flowing over the dam. Some snow. Quite a bit still on the ground from last week—about a half-inch. Fooled with water system again.

Carl got soaked under house. Found two or three more blown and cracked pipes. Gave up and put hose from pump through bathroom window into tub. At least we have *cold* running water. Very cold! But a big improvement over melting snow, as now we can wash dishes and flush toilet much more easily. We're just grateful when it doesn't freeze. Left water trickling overnight in tub.

Another cat showed up today. Hungry.

Hamburger the steer is getting very difficult—expects to be fed twice a day even though he has grass to eat again. Bellows ceaselessly.

December 29th, Saturday:
Sparky and Sylvester having standoffs on front porch. Howling and spitting. Hauled some wood out from barn in truck. Using fuel at a great rate these cold nights! Went for walk after lunch with Carl and Sparky and Sylvester, to bottom of property. Sylvester hung in there,but started to climb a tree when Hamburger followed him around (he doesn't fancy such big playmates).
Hamburger sets up a ruckus every time he sees us now. Stands outside our front door and bellows. Came back up a different route. All the little ponds are frozen. Carl walked on one, but I made him get off. Cats hissed at each other all the way down and back, but walked and trotted close by the whole time.
Fritz, far too old and dignified to have anything to do with such silliness, sunned herself on the deck while we were gone.
Made love on the sunny bed, then watched as the sun went down over the King Range.
Listened to Gregorian chants by candlelight and drank wine, watching stars come out. Had shrimp salad for dinner. Carl found a beautiful, fine bird's nest on our walk, entirely lined with cow hair, and a crinkly snakeskin.
Saw a few deer.

Monday, December 31st:
Today was better. The sun is out and it's warmer—almost 50°. Whoopee!
Carl put the big storm door on the front of the house and fixed yet another blown water pipe.

In the afternoon we walked to the top of the property, then across the road to Yewwood Creek and down to the riding ring (a relic of Marvin's early attempts to raise quarter horses on the ranch), then back through the woods and across little Alder Creek, and down to the barn by the lake.

Sylvester had to come with us, though I told him to stay home.

We had to carry him across the creeks and Carl put him in his backpack for awhile, but he didn't like that.

The silly cat went through an awful lot of snow up to his armpits, but hung in there until we got to the barns, then he climbed a warm rock and waited for us, delicately cleaning his feet. We arrived back as the sun was setting, and toasted the last sunset of 1990 from the deck. Still a lot of snow around. Lovely.

Tonight, we review the year, and look to the future.

\* \* \*

January 17th, 1991:

Got here two days ago, Tuesday (day of the UN deadline for Iraqi withdrawal from Kuwait).

Only two of the cows showed up. Went for walk and saw the others up on Yarborough's land.

We struck out armed with knapsacks full of hay and lured them back. Naughty girls had crawled under the fence! They came thundering down the hill with Big Mama in the lead, pushing Carl and butting him with her head (I was afraid she would step on his feet). Scary.

So good to have them back though. Give them hay and they'll follow you anywhere.

Got the water going yesterday, too. That is, Carl did.

I pruned the roses (what's left of them). Today we hiked down Yewwood Creek onto Walpole's property and discovered a spectacular waterfall.

We can pretend it's ours, I guess.

This afternoon we went to look at our waterfalls, as there has been a recent rain and a snowmelt combined with it, and the streams are running really swiftly. Even the little one we call Alder Creek by the corral has new pools and falls. We sat by the waterfalls under the Big Rock for quite a while, listening to the crashing water and watching it rush by. What joy!

But sadness, too. Yesterday, according to the *New York Times*, the Gulf War started, and though it's hard to imagine here, I am so worried and sad about it. I only hope it is over soon, with a minimum of suffering.

And here we are, amazed that others care enough about our way of life to defend it with their own.

* * *

When we returned to Zenia after the big dig, we went to the post office to pick up mail and visit.

We told the postmistress our harrowing tale of being snowed in for four days and having to dig our way out.

A neighbor who happened to overhear said, "Why didn't you call me? I have a tractor with a blade on the front. I could have had you out in 15 minutes!"

* * * * *

## CHAPTER 16

# Growing

*"And thou beholdest the earth blackened; then when
We send down water upon it, it quivers, and swells,
And puts forth herbs of every joyous kind."*
—The Koran, 22.5

Me and My Herbal Delights at the Christmas Crafts Faire,
Kettenpom

IT WAS ABOUT THIS TIME I HEARD ABOUT THE TRINITY
PROJECT.

Some of the ladies I met at one of the Fourth of July
celebrations had invited me to the Women's Farm Bureau
meetings, held once a month in the homes of members.

I decided at the time that it would be a good way to get to know my neighbors, so I managed to attend a few gatherings, but my attendance was sporadic.

Unlike the overpriced women's club luncheons in fancy hotels I had been accustomed to attending back in "the city," (by which I meant Santa Rosa), this one was always potluck. Everyone brought a side dish and the hostess provided the main course. We would have a brief business meeting to discuss farm affairs and current legislation, and then spend the next couple of hours eating, drinking coffee and visiting.

In a community where everyone lived so far apart, the Women's Farm Bureau meeting was one of the few times the women were able to get together and gossip. I expect our gathering was much like an old-fashioned quilting bee, with the possible exception of the political discussions.

We had no official status beyond the name, but that wasn't important.

What was important was finding out who needed help on her ranch, or who was ill or had died or moved, who had a cottage for rent, who needed a job, and most important, who needed a visit, a card, or a phone call of condolence or comfort.

At one such meeting, we heard from the county supervisor about her pet project: a proposal for federal funding to develop alternative industries for local folks.

The idea was that instead of harvesting timber or growing marijuana, people might find another lucrative (and legal) outlet for their talents. Besides, as the "underground economy" pays no taxes on their ill-gotten gains, the county was hurting for revenue.

Casting about for something we might be suited to do, she hit upon the herb business. After all, she reasoned, some folks already knew how to grow herbs successfully, so they ought to be able to compete in the then-infant U. S. medicinal herb business.

The rest of us had demonstrated some green thumb capability by managing to grow tomatoes at that altitude every year (some lived at 4,000 feet) despite a notoriously short growing season.

We all thought it was a great idea, but as is so often the case, that was simply because we didn't have enough information.

Nevertheless, I plunged in, starry-eyed and stupid.

I made a momentous decision, then. I would stay on the mountain while Carl took his business trips, look after the cattle and the cats, write, and grow herbs for the medicinal and ornamental market.

I'd give up my interior design business—or do it part-time by appointment—and devote myself to living off the land.

In anticipation of profits from growing my own herbs and the necessity of having enough to eat, I began to plant in earnest.

* * *

There was a slight problem. I would of course, be expected to attend monthly meetings at a central point for most of the county. This turned out to be a two-hour, 60-mile trek, through some of the most rugged terrain and over some of the worst roads in those mountains.

I soon realized that my monthly commute took me through some stunningly beautiful country, so it wasn't so bad a trip after all—at least until they started working on the road.

This project meant either delays of four hours or a detour of 18 miles each way.

In practical terms, if you had business in Hayfork, the town on the other side of the mountain, you had to leave home no later than 6 A.M. to get across the mountain before they closed the road at 8 A.M. Then you had to find somewhere to kill time (that meant the coffee shop or the hardware store—there was no library), until noon, when they opened the road again for an hour.

If you missed that time frame, you were stuck in town until 5 P.M., when the road crews called it a day.

I found I could only drink so many root beer floats, and there was no question of getting anything stronger before negotiating the mountain road again, with its serpentine curves and sheer cliffs, so I usually took a book and sat under a shady tree at the county fairgrounds and read the time away.

When the weather was good, this was a rather pleasant pastime, but in winter all that changed. As the snow and rain and freezing cold gripped the high country, attendance at meetings was sparse.

Our first meeting was an all-day affair.

People came from all over the county, and some had traveled even farther than I had to get there. About 100 families were in attendance at that first meeting.

I looked around the room at my fellow cooperative members. It was a very mixed group: mountain men and aging hippies (often indistinguishable); nouveau back-to-nature types in brilliant tie-dye, flowing hair and beads; eager young people with muscular arms and strong backs; middle-aged folks like me; students, and even a grandmother or two.

A few of them even looked like they might have done a little farming, but most bore the same rapt and zealous demeanor I did.

The meeting itself (a potluck, of course), turned out to be one of the most interesting and informative I had ever attended.

The keynote speaker was a professor of earth science and a commercial raspberry grower, and he was an expert in organic farming practices and marketing. I was enthralled. He spoke for six hours, and I could have listened six more if the seats hadn't been so hard.

But best of all, I came away from that meeting with plants—real live ones, and they were FREE!

I lost no time in planting mine, but then communications from the other side of the mountain were sketchy, and I was still commuting to the city nearly weekly, so that first year I just tended to my new charges and wondered what to do next.

* * * * *

## CHAPTER 17

# Letter From Zenia

May 15, 1994

Good news! High Mountain Herbs, our local Southern Trinity Herb Cooperative, has been awarded $64,000 in Option 9 funds to provide jobs and business start-up costs as we grow toward commercial viability in the herb market.

The money is to be used for crews to assist farmers with fencing and planting, organic certification and office equipment.

Cloud Lake Herb Farm, my own business, grows slowly as I hobble around on painful feet, staking and planting and weeding and propagating. I look forward to next year when I will stand on my own two (repaired) feet and harvest great armfuls of fresh and fragrant dried herbs for potpourri, for cooking and for sale.

This is the best growing area I've ever had to work with—plenty of water and sunshine, and the sort of fast-draining soil herbs love, so, except for those times when the cows break into my garden and ravage everything green in sight, I feel, like Doctor Pangloss, that I have landed in a gardener's Best of All Possible Worlds.

May your garden grow as well as mine.

May 16th, Sunday:

Put in three grueling hours this morning, hauling hay and planting potatoes. Finally got hold of Cindy, and she came down to find out how much to feed the steer while I'm gone. Nice lady who has had a hard life. Her husband seemed so sweet, standing there on the porch with his toes peeking out of his socks, saying "Y'all can come here any time."

Kids are cute too; they love it here. Hope they can stay awhile.

Broke down on the phone tonight with Carl, who is out of town on business again.

So sad about the steer going to market. He's really getting to me. Cisco.

May 17, Monday

Snazzy had half my bottle of fish emulsion fertilizer for breakfast. Chewed off the cap and the label and helped herself. Not even a tummy ache.

Planted Oregon Sugar Pod peas, and the rest of tomatoes (Oregon Spring) that I raised from seed, 10 plants in all. Also planted dill and Italian parsley. Got water to garden, but automatic system doesn't work. Will try again to fix it tomorrow.

Fertilized flower garden. Wrote. Read.

May 18, Tuesday

Planted my thyme in last row. About 30 plants, maybe more. If nothing else, it'll dry well for wreaths, etcetera.

Sprayed pasture area near houses (again) for thistles. Killed a little poison oak within our newly-fenced compound. Sprayed three gallons in all (three gallons is <u>heavy</u>!)

After lunch, I worked in the garden some more, took a shower and got ready to go to an herb co-op meeting in Van Duzen at Phoebe's. She's 75 and still going strong. Got some licorice mint and tansy seed from her.

Still nothing from TAB (Trinity Alps Botanicals, our parent organization), and here it is nearly the end of May! What a screwed-up mess.

Already we are mired in governmental bureaucracy!

Snazzy went with me on the herb meeting trip. I was fearful about it—so lonely on that road over Grizzly Mountain at night. All my childhood fears of dark forests came up again.

It was comforting to have the dog, but she, of course, had to bark loudly into my right ear and fling herself at the car window every time we saw a deer, bunny, toad, or mouse, or any time we crossed a cattle grate. As the road through the deserted national forest is mostly open range, we crossed quite a few. This did not help my nerves.

Met my first gravedigger at the meeting last night—Tony.

Transplanted to the city, he might pass as a street person, but this guy works *hard* for a living. The dirt from the graves he digs still clings to his clothes and lives under his fingernails, as though he can't wash it out. His girlfriend works for the court, which she says will close soon due to lack of funds, leaving the only court in the area in Eureka, in Humboldt County. She thinks the sheriff's office will be forced to close, too. Hard times.

\* \* \*

May 26th, Wednesday

Spent the morning spreading hay over my newly raked beds and planting cilantro, dill, cinnamon basil, parsley, peppers, Scarlet Runner pole beans, lettuce, and cucumbers. Snazzy helped.

Started raining by late afternoon. A happy discovery: KMUD (our local—and only—FM station with varied fare) has jazz from 4 to 5:30 on Wednesday. Better than the usual political tirades. Soothing on a rainy afternoon.

May 27th, Thursday

It has rained all day and most of the night. We had 2" of rain in one day. My plants are drowning!

Steer fattening nicely. I'm valiantly trying to eat up last year's produce and meat in freezer to make way for the new stuff. My seeds should be up tomorrow. If it's clear, I'll plant licorice mint and tansy. Meantime, I clean my office and try to get organized. Dog extremely bored.

May 28th, Friday

Wet and windy, but not as cold. Had 3" of rain in the last four days. Yesterday I did some writing and a little background reading, and not much else.

Today I discovered the steer had broken the bottom rail off a section of the corral trying to get to the green grass on the other side, so, after

I fed him, I trekked back up the hill for hammer and nails and fixed it. Then I added some wire to the bottom of the garden gate, so small rodents can't get in. Of course, there's no stopping the gophers.

Cows all watched this activity as they relaxed in the big pasture. (Such a lot of fuss. People are so strange.)

Made manure tea for my plants in the big barrel. Put it under the drip line of the barn roof—worked great.

This afternoon I sieved potting soil and planted flats of tansy and licorice mint, plus purple ruffles basil, feverfew, pyrethrum and lavender Munstead. Transplanted Thai basil.

Veggie garden report: peas are up, beans just starting, onions up. The rest appear to be in an underground holding pattern until the nighttime temperatures in these mountains warm a bit more.

May 30th, Sunday

Rained all day—steadily and hard. Roof is leaking all around me. Snazzy is bored. Took a day off and did some heavy reading. Had a nice fire and was quite cozy. Dog napped and played with her toys, but stayed close by.

May not be able to sleep in this bed tonight—ceiling is starting to drip in some strategic spots. Tomorrow's forecast is for "showers." Ninety percent! Also Tuesday and Wednesday. Maybe some on Thursday. Then more rain Friday.

Time to build ourselves an ark.

\* \* \* \* \*

## CHAPTER 18

# FOLKS

*"There were never in the world two opinions alike,
any more than two hairs or two grains. Their most universal
quality is diversity."*
—Montaigne

THERE WAS A TIME WHEN WE WOULD HAVE BEEN perhaps more judgmental when confronted with the type of "diversity" our little community offered, but that was before we knew the folks behind the labels: pot growers, retirees, yuppies, loggers, aging flower children, artists and other assorted Bohemian types, grannies, Indians, mountain men, the college-educated and those who did not finish grade school, cattle ranchers, L.A. fugitives and lifelong residents all seemed to fit together when they needed to, only to split and re-form in yet another configuration as circumstances dictated.

Such was the nature of expediency where our little patch of mountain greenery was concerned. And we were not unique. As anyone who has ever lived in a small town knows, we are all interconnected and interdependent and, in the country, you must understand this simple principle because your very life may depend on it.

This is not to say we all liked each other—far from it. People who thought alike always found each other just as they do in the wider world, and they always managed to heap scorn on those who thought differently.

Gossip was always appreciated and the art was zealously practiced, not just by the women, but by everyone in the community. Especially for those who had no electricity, it was often their only source of news.

But if your house burned down or you got laid off or someone close to you died or the Community Hall needed repairs, you'd find a strange assortment of people working side by side and thinking nothing of such mingling as you'd be hard-pressed to encounter in the city.

So it was with no surprise that I found myself one potluck evening at the Community Hall, sharing a table with Chicken Charlie.

As I've said, during the time we lived in Zenia, our joint community of Hoaglin Valley, Zenia and Kettenpom boasted about 250 registered voters. This was not an entirely accurate way of judging the population of our small part of Trinity County, which only counted about four people per square mile within its nearly 3,200 square miles, partly due to its inaccessibility and partly due to the fact of all that government-owned land—and partly to the lack of jobs in the area.

There were those who neither voted nor paid taxes; who lived off the grid and tried their best to convince the rest of us they didn't exist, except, of course, when we stumbled into their illicit forest gardens—an activity that was known locally to be exceedingly dangerous, if not fatal.

(The local folklore told of trespassers being shot or at least threatened with loaded weapons, people stepping on bear traps or land mines—land mines?— or blundering into fish hooks, poised at eye-level, and other colorful stories.)

Such a one was Chicken Charlie. This was not his real name of course, but I will always remember him that way.

A relatively new addition to our group—if, in fact, such an odd assortment of characters could be called a group—Charlie was tired of living alone on the edge of creation, and craved some human interaction. Thus, he had ventured out on that particular rainy, miserable winter night to share a meal with the assemblage and warm himself at the Community Hall woodstove and in the combined warmth of people and their gossip.

He sat down opposite me, his paper plate so heaped with food it threatened collapse, sighed contentedly, and attacked the tuna casserole.

Charlie was a colorful fellow: he had one long, thick braid from which straggled strands of yellow-brown hair that looked as if it hadn't seen either a hairbrush or a wash in some time; dusty, well-worn jeans; and a brilliantly tie-dyed undershirt that peeked out through the holes in his old blue denim shirt. A jaunty gold earring adorned his left ear.

Finally, after a protracted silence and a somewhat reluctant exchange of names, I ventured a question: "How long have you lived on the mountain?"

Charlie brightened. "Been here about six months now. Got me a little place over to Covelo way-down off the old road."

I asked a more delicate question: "What do you do there, Charlie?"

"Ain't quite decided. Gonna get me some chickens. Gonna raise me some chickens, and then I'll see. Thinkin' of carvin' me some totem poles. Sell 'em at the county fair. Ain't quite decided though—hafta see what happens with them chickens. Might sell me some eggs an' some fryers. Just hafta see."

Fearing that his career choice might also be somewhat dependent on the pot crop that year, I decided to probe no further. We shook hands, said "Pleased to meet you," and I wandered off in search of more food and less veiled conversation.

\* \* \*

Everyone on the mountain was multi-talented. The local herbalist made candles (or did, until her house burned down), replanted forests, cleaned houses, and sang and played guitar in the local country band.

Our resident artists usually sustained themselves with waitress work between infrequent sales of paintings, and the cowboys were as itinerant as cowboys historically were, traveling wherever the work took them, often leaving their families far behind.

In that sense, I fit right in, propping up my interior design business with frequent trips to town for consultations—there was no market for interior design in the mountains—making flavored oils I sold to the Peninsula Hotel, substitute-teaching and growing organic herbs.

In this latter pursuit, I was as happy as a cow loose in the hay barn, nurturing my seeds, planting my cuttings, and constructing wreaths and potpourri and fancy vinegars for the annual pre-Christmas bazaar at the Kettenpom Store.

When the occasional CAMP (Campaign Against Marijuana Production) helicopter flew low over my herb garden, I'd wave and smile, and they would go away.

I should explain that living in this remote part of Northern California is not at all like living in rural New England or Pennsylvania or upstate New York, their landscapes dotted with picturesque farms encircled by fieldstone walls, quaintly weathered barns and split-rail fences.

Rather, it is the Western Frontier, with rugged and individualistic people, and if something is not functional, it does not deserve to just lie around, cluttering up the place. Form is replaced almost entirely by function. It has to be, as there is no room for anything else where survival is the order of the day, and except for a scattering of writers and artists, attracted by cheap land and the unspoiled beauty of the place, everyone there made his living one way or another off the land.

One of our older neighbors had his livelihood all figured out, until he was kneecapped by governmental regulation.

"I've only got 160 acres (a small parcel for this area, known as a "square"—four adjacent 40-acre parcels), but on it I've got some prime old-growth timber. Of course, I wouldn't want to log it all, but I expected to take a few trees a year until my second-growth fir is ready. That's my retirement plan," he said.

But, of course, he couldn't do that. The latest government regulation (known as the Spotted Owl Decision) had dealt a crippling blow to the already-sagging local economy, and Trinity County—one of the poorest in the state—was one of the hardest hit in the whole Northwest.

Living there, one got to see both sides of the controversy, and either way, it was not pretty.

* * *

In such a small community as ours, deaths or defections left large holes.

Our dearest friends from the mountain were no longer there, though their presence pervaded our home.

But Marvin and Nancy Scott, the former owners of "the Scott Place" have remained close friends, both during and after our Twin Creeks experience.

Our relationship was unique because when we met them on our first trip to the ranch, we felt immediately as if we'd known them all our lives. Indeed, they shared with us many stories of their fond memories of the place before we had even moved in.

As easily as Carl and Marvin bonded, so, too, did sweet Nancy and I.

That first afternoon, we passed the time waiting for the menfolk, talking about ourselves, our children and our dreams. She was a dedicated gardener, too, so of course we swapped gardening stories. Nancy regaled me with tales of the amazing fecundity of that 120 acres, poised as it was between the snow line and the valley and blessed with abundant water, friable soil and sunshine.

And of course, there was a handy source of fertilizer as near as the barn.

I was ready to grab a shovel by the time she was done.

Nancy and Marvin had five children, now grown and with children of their own. Unfortunately, their offspring were not as enthusiastic about the ranch as they were.

It was too remote, too difficult to visit, and there were dangerous snakes their children kept dragging home, along with poison oak, mosquito bites, etcetera, etcetera, etcetera.

Mainly though, I think the Scott children worried about the health of their parents.

For one thing, Marvin had a heart condition and, at 72—after his good friend, a Burgess, died on the way to the hospital after a heart attack—was finally persuaded it was time to take life a little easier (though now in his 80s, he has yet to do that).

Marvin and Nancy had started their ranch life by raising quarter horses, moving up 10 years earlier from the East Bay, where Marvin, a former military pilot, had been the manager of Buchanan Field, a large airport in Concord, California, which catered mainly to private pilots.

They found out quickly that raising horses was more labor-intensive than they had intended, so they switched to raising Polled Hereford cows. That was the beginning of a long love affair with those animals that continues to this day, if only in memory.

Whenever they called, Nancy never failed to ask about her roses and the fruit trees they planted, and Marvin would always ask about his redwood trees, and they both asked after their beloved cows.

Marvin was as driven and intense as Carl.

When they made their first move away from the

ranch and closer to the hospital, the first thing Marvin did was to plant 100 redwood trees, plow up the sunny flatland for a sizeable vegetable garden, plant fruit trees and install a five-tank water system. Whenever we visited, he always had some new projects to show us.

Both should live to be at least 100, and we sincerely hope they do, for there are no more generous, thoughtful, friendly and good-hearted people on the planet than these two. We have been proud to call them friends.

\* \* \*

And then there were the woodcutters.

We were almost completely dependent on wood for heat at the ranch, and though we did cheat with electric space heaters in the darkest winter days, they were expensive to operate, so we used them sparingly.

This meant that we either had to cut and split our own recently downed wood (which, as some wise person noted, warms you three times—once when you cut the wood, once when you stack it, and once when you burn it), or depend upon an ever-changing cast of characters to provide it.

Busy as we were, juggling two distinctly different lifestyles, we generally chose the latter solution to our heating problem.

And that is how we came to meet "The Shifter."

Usually, we would call on the woodcutter we had done business with previously, as Carl was assured by careful measurement that we actually got the four or five cords of wood we paid for (something of a problem in a place where one was considered a greenhorn, and therefore subject to the delivery of lightish loads).

Each year we would leave a message at the general store for the previous year's woodcutter, and each year we would get the same message as soon as the woodcutter surfaced: "Gee I'm sorry, but I can't help you this year. I've got a lotta orders to fill, and it's been so wet I can't get into the woodlot to cut any more. Anyways, I expect you're not after green wood."

Who else would he suggest we call, we would ask with mounting anxiety as we contemplated a woodless winter.

Sometimes we'd get a lead that way, but it was usually quicker to mention our projected heating problem at the post office or the general store and then wait for the bush telegraph to work; but that year we were late in asking.

The other factor raising our anxiety level was the fact that the longer we had to wait for wood delivery, the higher the price would be, and the greener the wood.

That particular year, the luck of the draw yielded up Jim Stover and his 13-year-old son, Doug.

They showed up one particularly frosty morning in a dilapidated Chevy truck. The dog, generally a reliable doorbell, did not announce them, being at that moment immersed in a satisfying dream of chasing rabbits, or so we interpreted the rapid paddling of her feet as she lay sprawled on the doorstep, eyes closed, twitching and whimpering.

No, they were rather announced by the rude grinding of gears as the ancient pickup made its way down the steep road to the ranch house.

By the time we had reached the front step and negotiated our way around the sleeping dog, Jim was vigorously applying the hand brake, as the foot brake was apparently not too reliable, even on the flat drive in front of our garage.

Doug jumped out of the truck and immediately busied himself with the tailgate, as Jim attempted to extricate himself from the vehicle, extending a burnished hand in greeting.

"This here's my boy," said Jim with a touch of pride. "He's gonna help me unload today. Doug, come on over here and meet the folks."

Doug continued to hang back, but he waved and smiled weakly.

"He's a young'un, but he's my right-hand man," Jim said. "Don't know what I'd do without 'im. Where'll you be wantin' this wood?"

"I got a place all set up for you down in the barn," Carl said, indicating yet another steep downgrade to be negotiated by the old truck. "You think you can make it down there?" he asked dubiously.

" Oh sure," said Jim. "I got me a secret weapon I call 'The Shifter.'"

" Oh yes?" said Carl.

"Yep. Allow me to demonstrate," Jim said as he got back into the cab and yelled back to Doug. "Put that tailgate back up and get in here. We got us some shiftin' to do."

Without delay, the boy jumped into the cab and disappeared under the seat.

We were mystified at what happened next.

Grasping an extremely truncated gearshift where it sat below the floorboards in a large hole in the floor of the cab, Doug deftly shifted the truck into low gear for the descent, as his father worked the clutch, gas and hand brake.

"Ain't he somethin'," yelled Jim proudly as they made the perilous trip down the hill to the barns.

We puffed our way down after them on foot, just in time to witness their arrival.

"Dang gearshift broke last month, and I've been too busy splittin' wood to get it fixed," he explained. "The kid's pretty fast though, so we worked it out between us. Where you say you want this wood now?"

Carl managed to point to the barn, but words failed him.

Nevertheless we were able to see the whole process repeated in reverse as they wended their way up the hill and out of our lives.

Carl was laughing so hard he couldn't stop. I think he overpaid Jim, but it was worth the price for entertainment alone.

It reminded me again, that mountain folks possess resourcefulness that borders on cunning. They have to, to survive.

\* \* \*

CHAPTER 19

# Mountain Women

*"Something was always going on in the corral, and I would leave the dishes standing in the kitchen and run down and watch, sometimes for hours."*
—from *A Bride Goes West*, by Nannie T. Alderson and Helena Huntington Smith,
University of Nebraska Press ©1942

SOME OF THE WOMEN I ENCOUNTERED ON OUR mountain did have help, especially on the big cattle ranches. But most were like me—with just the two of us plus an occasional hired hand, it wasn't a matter of watching what went on in the corral, but rather one of working *in* the corral out of necessity, doing things I'd never, in my former life, dreamed I'd be capable of doing.

In that sense, it was challenging, affirming and rewarding.

In another, it was frightening, dirty and exhausting.

The womenfolk were split between those who had always lived there and knew what they were doing, and those who had moved from the city and were, like me, frantically trying to learn how to live on a ranch. Even *they* had more experience than I did.

There was Joan Barnes, the local amateur historian and hostess to many, many Women's Farm Bureau meetings, for one thing because she had a deck large enough to accommodate as many as 25 women at one time, as well as a motley assortment of coffee mugs—enough for the diehard coffee-drinkers in our group.

Joan and her good friend, Samantha Smith, made things happen around the small community, increasingly with help from some of the younger members. Samantha was a transplant from Los Angeles, and had taught there before she retired, but she adapted to country life easily, and gave of her time to the community with unstinting ardor.

Ella Chambers, one of our near neighbors (and none of them was very near), slender and silver-haired, with an unexpectedly regal bearing, was always a source of high-energy goodwill, free plants, free food, and a variety of crafts.

The quilts she made were much sought-after, and when she died, in her 80s, she left a very large hole in the community.

Jeri White, who was for several years our ranch manager, and, herself, a great addition to the group, recalls a wonderful Ella story:

"Ella and I were driving together to a Women's Farm Bureau meeting, and when I got to her house, she answered the door in a giddy mood. 'Guess what I'm bringing?' she said gaily."

"I give up," I said.

"'I'm bringing a cake. It's called, *Better Than Sex Cake!*' she said, blushing." Ella was a very proper lady, and she hadn't uttered the word, 'sex' in public very many times.

"I laughed. 'Well,' I said, 'Tell me, quick, what's in it?'"

"She blushed again, and said, 'It's white cake mix, with layers of chocolate, cherry pie filling, Cool Whip topping (a staple on the mountain), and nuts! I just *had* to try it!'"

"Ella bore her cake triumphantly to the meeting and before she was through, she had shyly and proudly made the rounds, telling everyone the ingredients, but especially the name of the cake. All present had a slice of her *Better Than Sex Cake*, and both Ella and her cake were a hit that day; but it was way too sweet for me."

Now that she's gone, we miss Ella's *own* sweetness most of all.

* * *

When Sylvia Horton and her husband, Paul, moved to the mountain from the city full-time, they started our little community paper, "The Hoaglin Zenia Community Newsletter. (Samantha was one of the instigators of that project, too.)

Aside from the newsletter, all our news came from Humboldt County, from the *Life and Times* (weekly newspaper), and the local FM station, KMUD, presided over by the estimable Estelle Fennel, faithfully, breathlessly and tirelessly reporting each new breach of trust or outrage on the part of Pacific Lumber—which was at that time logging in the Headwaters' ancient redwood groves. Estelle covered the exploits of the Campaign Against Marijuana Production and the Environmental Protection Agency, as well as which protestors had been pepper-sprayed by the police, who had been arrested and jailed and for what (spiking redwood trees was popular), and the current state of the Peace Movement.

Aside from the American Indian Report, nothing else appeared to be news.

Newspapers and newsletters saved from 1994 through 1999 reveal the shocking depth of unrest in our quiet community. And we were foolish enough to think we were getting away from it all!

A selection of headlines from that time:

From *The Sonoma County Independent*, 1998
**Virgin Sacrifice**: Touted as the greatest piece of environmental law in decades, the Headwaters deal gives away the forest to save the trees.

From *The Press Democrat,* 1998
**80 Days in a Tree**: Earth First! protest gathering attention.

From *The Independent News of Southern Humboldt,* 1999
**Federal Commission Hears Testimony to Decide Fate of Eel River Water:** While the Eel River was historically the third largest fishery in California, it is now entirely closed to commercial salmon fishing.

From *The Press Democrat,* circa 1999
**Eel River Exports:** EPA jumped the gun in encouraging an end to diversions to Russian River.

From *The Southern Humboldt Life and Times,* 1999
**Neighborhood Upset by Cell Phone Tower:** The neighborhood on Lower Sawmill Road, just off of Alderpoint Road, has an unwelcome visitor.

From *The Sonoma County Independent,* 2000
**The Last Tree Hugger:** Environmental activist Julia Butterfly comes down to earth. And so on.

So much for getting away from it all!

\* \* \*


For a while, I tried to be part of the solution by becoming a member of the Hoaglin Zenia Workgroup, an arm of the Southern Trinity CERT, or Community Economic Revitalization Team.

This was a group formed with what was known as the Option 9 Grant, in which federal government agencies, on the advice of President Clinton, agreed to help "economically depressed areas" such as ours with jobs, business loans, training, technical assistance, reduced red tape, and even—

when absolutely necessary—money in the form of government grants.

During that time, I am listed in the Hoaglin Zenia
Community Newsletter as the chairperson of this august
group, though I can't recall chairing anything. I do remember attending an awful lot of meetings in Mad River, however (20 miles away over a lonely mountain road), as the representative from our group.

The idea was to come up with our own solutions to
the vast unemployment problem that had been created in our
community by the aforementioned endangered species list-
ings, thus rendering most of the old-growth forest off limits
to logging, which, aside from the underground economy, was
the chief employer of our community's able-bodied workers.

We thought of developing businesses for hunting
guides, some sort of tourist bureau, fish camps, word-pro-
cessing and Web-design businesses, and a lavender distillery,
which was to be a part of High Mountain Herbs, a local off-
shoot of the Trinity Project. All of these, of course, would
require training, especially as few of us at that time were
computer-literate.

We also thought a multi-purpose community center
to replace the old Grange Hall might be appropriate.

We were appointed by the CAG (Community Action
Group) to submit our ideas to the EDC (Economic
Development Council) and wait patiently for government
money to rain down upon us—I do remember that some of
the more seasoned members of our committee were somewhat dubious about the occurrence of this latter event—
whereupon our OEDP (Overall Economic Development
Plan) would be used to guide the revision of the CGP
(County General Plan).

It was at about this point that my eyes began to glaze
over, and I suddenly found I had other pressing business to
attend to. Nevertheless, I remained on the mailing list for the

next four years, during which time, as I recall, not a lot happened, besides a tremendous waste of paper, and therefore, trees.

So much for reduced red tape!

As Carl was fond of saying, "That's what happens when you let the government into your plans." He had a point.

At least our side of the mountain did manage to get itself a couple of computers and a medical clinic, and train EMTs for medical emergencies, and bloodhounds for search and rescue—two components of our infrastructure that were sorely needed. For a while, we even had a doctor in residence. And we did get a couple of computers, which were housed in the office of the Kettenpom General Store. We got a printer, too. All of these had to be spoken for on a rotating basis with a weekly sign-up sheet. The office was staffed by volunteers, but was recognized as our local government branch office. It was often crowded.

I gave up on that and used my old, slow Mac at home.

Internet access was either dial-up or nonexistent.

I remember we were also working on getting some kind of quick-response fire service, as the forestry service fire trucks—a half-hour away over mountain roads—tended to arrive after the fires were out.

I know, because I had to call them when a neighbor accidentally set fire to her fallow summer field, and we had to rush to wet her house down while at the same time trying to contain the grass fire, which was out of control and moving rapidly toward the house.

By the time CDF (California Department of Forestry) arrived, everything was fine. But at least the firemen were cute.

But back to the women. They are a strong breed, these mountain ladies, and able to handle whatever life throws them as challenges to be cheerfully overcome—

whether out-of-work and underfoot spouses, husbands who do a little more drinking than they should, fences that need mending, stray cattle, mountain lions, rabid dogs, rabid skunks, sick children, hungry logging crews, wayward cows, or breech births.

I stand in awe of these women, and I salute them—heroines, every one.

* * * * *

## CHAPTER 20

# Letter From Zenia

MAY 3rd, 1996

Came up in time to see a lovely sunset on the way. Dead mouse in living room. Hoards of mouse poop and poison among the drawers of clothing. Welcome home.

Wasps at least are somewhat quiescent, but we arrived ready to do battle with them as well. Lots to do today, but will try to go about it in a mindful way: weeding, watering, mowing, cleaning house, evicting mice, washing clothes, bedding, etcetera, etcetera.

Cows in the pasture, sedately roaming and selectively munching. The babies from last year are quite large by now. How quickly they grow. We're about to go up to the orchard to visit our fruit trees and to see if there are any late lilacs to be had.

The lilac bushes are finally tall enough that the deer don't bother them much, so I am hopeful.

So good to be here and just to be outside. Even to do housework. It occurred to me this morning as I meditated by the fire that the big secret of life is that any task done mindfully is good and rewarding—I knew that. Just had to remind myself.

Ken and Sarah from Forestville are coming

up tonight for a couple of days, so we are in vegetarian mode.

Snazzy is happy as a dog can be, chasing everything in sight. I long to chase things, too.

Carl happy today. Had to fix water system, broken again when we arrived last night, but whistled the whole time he worked.

Saturday May 4, 1996

Didn't succeed in being mindful yesterday. I washed clothes and bedding, cleaned house and banished the mice, and I went at it pretty hard. Today we will walk to the falls.

Our guests are sleeping peacefully in the unpainted bedroom. They brought some nice wine—an organic chardonnay—and we enjoyed that as we watched the sunset, seated at the burl table in the living room.

The ranch is so lovely this time of year. Wildflowers are late, but they're just getting started now. Goldfields, meadowfoam, poppies and lupine, and lots of small stuff like cranesbill, dogtooth violets, Indian pinks and buttercups.

Dear God, it's good to be here.

9:30 P.M.

Carl on phone with ex-wife, Nuri. Sarah and Ken reading. Dog after big game in the attic.

There seems to be a raccoon up there. Carl and Ken lifted dog into the attic from the bedroom to see if she flushed anything out.

No luck, but she's still pretty wired, so she must have seen the critter.

Today Carl, Ken and Sarah hiked to the top of Big Rock. I waved, then sat atop the dam spillway and watched the fish. It was so peaceful there, but I kept looking over my shoulder for mountain lions, which are rumored to be in the area, so it was not altogether tranquil.

\* \* \*

Sunday, May 5th

Up at 6:30. Troops up at 7. Waffles for breakfast. We've decided there may be a colony of bats living in our attic, over the bedroom window where the wasps come in. There seem to be considerably fewer wasps all of a sudden, and we have seen bats flying around past the window recently. I hope so.

We need them.

\* \* \*

June 5th, 1996

Morning sounds. Dense layers of birdsong filling the air from far away to very near. Carl coughing, spitting, blowing his nose, shaving—very, very near. A full range of sounds, from sublime to silly.

Cows will be arriving soon from wherever they spent the night, mooing and jostling and shoving and stomping around the barn. I hear them in the distance, calling to their calves.

Gorgeous, sunny, temperate day. No wonder the birds are happy.

Carl and Tim are going to put the cows through their annual agony today: ear tags, shots, castrating, branding—the works. I am unable to do it again this year because of my feet (some inflammatory problem modern medicine can't fix). Tim helped last year, and our neighbor, Cindy, did it the year before. Thank God for handy children, good friends and neighbors.

Snazzy spent the night outside last night, in disgrace. She, who has been sprayed by a skunk at least twice before, has proven herself a slow and stubborn learner.

Carl and Tim took her down to the lake last night, and she flushed out a skunk from underneath the bridge. The skunk started to leave and the stupid dog chased it, getting the full treatment right in the old chops. One can only hope this has made a permanent impression on her memory. I'm the one who has to haul her back to town tomorrow. Perhaps I should pray for plugged sinuses!

June 6, 1996

Another gorgeous day in paradise. Promises to be hot, but maybe we'll get lucky and have some afternoon wind again. That would be nice. Supposed to be in the 90s in Santa Rosa, so I think I'll dally here awhile before I try to make it back to town tonight. Got to do battle with the weeds again. Try to get everything watered before I take off.

Tim and Carl mended fences yesterday. They're gradually working up to doing the deed with the cows. Had some of neighbor Harvey's ham for supper last night. Dry, but good flavor. I made peach and blueberry ice cream, and we have some more cookies that I made the night before last.

Bring on the lovely day.

* * *

October 4th, 1996

Bulletin from KMUD FM: Acacia reports from jail that the Pacific Lumber folks have been spreading deer and fish guts around under the trees that comprise the "Ewok Village," the temporary homes of the tree-sitting species of timber activists who are currently camped out in the Headwaters National Forest, trying with their bodies to stop the slaughter of ancient redwoods.

Estelle Fennel also reported that a Mr. Barnaby was arrested near Garberville for shooting at his neighbors, who were doing repairs.

"This is not the way to settle disputes, with a 12-gauge," said the sheriff. Indeed.

She reports another incident involving two drunks near Honeydew, one of whom shot out a power line, frying computers and microwaves in the area. An acquaintance of the shooter was heard to remark, "If he'd been *aimin'* at it, he wouldnta' *hit* it."

Got home last night about 8 P.M. Unpacked, turned on water (tanks both full, thank God), and sat on the back deck watching the stars. As there was no moon, it was easy to see trillions.

Up at 7:30 and down to look at our new babies, to determine if any were males. Discovered we had only one castration ring, and started to worry. Rounded them up and discover they were all females—what luck!

Lily looks pregnant—still—hard to tell, but we might have one more. That would make 17— our biggest herd yet.

Spent the rest of the morning trimming up the big pine in front of the bedroom, hauling the branches into piles for burning; watering; collecting fir cones; putting flea powder on the dog.

Ate lunch on the deck. It's about 70° and cloudy, with wind. Very nice, except for the wind.

Calves gamboling in the field this evening— jumping and running. Butting each other. Drove to top of property to see sunset.

October 27th, Sunday

Waffles for breakfast. Back on Standard Time, which means early darkness. Cozy though, as it's decidedly colder at night—28° yesterday morning on the front porch. This morning, it's sunny and breezy.

Carl fed cows in the morning and they stood around, bellowing, the rest of the day. We went out to work on the roof in the afternoon, and they stood around in the pasture, bellowing. Each time I tried to talk to Carl or he to me, it set off a chorus of bellowing. Pretty damned annoying, that. He's gone down to feed them now in the hope of producing satiation and a little peace. Really windy today, but supposed to be warmer.

Our strategy today is much the same as yesterday's: stay inside in the morning while it's cool, then up onto the roof when it's warm enough for the roofing material to flow.

We're still on step one—filling all the cracks, seams, and nail holes. Next, we have to roll white latex coating on the whole roof.

That part comes later.

For now, if we can even stop the roof from leaking over our bed, we'll be happy. (We had mushrooms growing in the bedroom carpet last winter!)

Hawks and crows busy this time of year. They must be preparing for winter, too. I could hear the hawks calling to each other all afternoon from my perch on the roof. Felt a little birdlike myself... but no flying yet, except in my dreams.

Wrapped rope around my waist in case I fell, but I kept getting tangled up in it and dragging it through the freshly applied tar. Messy. Messy *and* precarious.

The other thing we have to remember is to build the fire in the woodstove in the morning and let it die down at noon, so we don't get-gassed or smoked out when working so close to the chimney.

After I finished yesterday, I harvested sage (three screens-full from one plant!) and cut back the lemon balm, artemisia and tansy.

Never did get any tansy flowers last year. Deer ate it all.

Last night after supper we had a bubble bath and lay in front of the fire, holding each other and listening to the Modern Jazz Quartet (MJQ, for those who remember them) by candlelight. A good day.

Carl out yesterday morning patrolling the property to ward off intruders.

Signs of same down by the dam—cigarette butts and shotgun shells. People and dogs apparently killed one of our new calves. So sad. Lilly's firstborn. She looks so forlorn.

It is the last weekend of deer season. Thank God!

* * *

December 7th, 1996

Cows glad to see us when we got back from town. Bellowed in unison as long as we were within sight of them. Had six inches of rain or more while we were gone, so everything is turning green again.

Did a little house cleaning and harvested more sage and rosemary. Will give some away for Christmas.

I got up early and stood in the rain in my oversized yellow slicker to hand Carl tools for repair job. He was again crawling on his belly in the mud to fix a leaky pipe under the house.

December 9th, 1996

Raining. Hard. Carl looking for another tiny piece of paper amid the flurry of tiny papers that orbit him and the dining table. Much frustration. Best to stay out of the way.

Built fire and climbed back into bed with my morning cup of coffee.

Worked all day, feeding cows, unloading hay we bought from a nice old rancher in Laytonville. He showed us the house he'd built for his wife.

"She won't stay here," he said, sadly. Seems she prefers Sacramento, despite the lovely house on a rushing stream. The old guy was really lonely.

December 10th

Loaded firewood and our freshly cut Douglas Fir Christmas tree onto the truck—tree was so heavy, and was 20' long—and drove back to town without getting stuck in the mud. Quite a feat, as we've already had at least 16 to 20" of rain this year at Twin Creeks Ranch.

* * * * *

CHAPTER 21

# DUST NEVER SLEEPS

*"Well, well, well—here I am at Twin Creeks Ranch, and I am so ecstatic and happy…Had to keep reminding myself it is OK to have days like this. Especially with where my mind and body have been for all these months."*
—Jeri White, *Backdrop for Joy: Lessons in Shadow and Light (The journey of a mother's heart after the death of her daughter…)*

*"Where there is sorrow there is holy ground."*
—Oscar Wilde, De Profundis

AS IT TURNED OUT, ANOTHER STRONG, AMAZING woman was about to enter our lives...

We were drowning in housework, yardwork (my fault) and rancher's responsibility. Finally, we decided we needed help.

Jeri came to us in response to an ad that, in our desperation, we had placed in a publication called *The Caretakers' Journal*, which existed for the purpose of matching potential caretakers with their clients.

Just what we needed.

Her entry into our lives turned out to be a miraculously fortunate event for all three of us.

Jeri had short, naturally blonde hair, short fingernails, and a sweet, open face lined with sadness. She was ten years younger and nearly a half-foot shorter than we were, and I worried she might not have enough stature at around 5'5" to intimidate a cow, but as we got to know her and her gentle ways, we found she possessed hidden reserves of power.

We had three responses to our ad. The other two were from cool, intellectual types (also single women) who were "in transition"—I thought that was a given—and looking for a retreat for themselves, which, they said, should not involve too much work—after all, each of them was awfully busy writing the Great American Novel—and it should not be too far from San Francisco, so they could get their culture fix on a regular basis.

Only Jeri sounded sincerely interested in the adventure of the isolated life in the wilderness we were offering.

We met on neutral territory, at my mother's retirement complex, as I was spending my weekends with Mom and commuting to the ranch during the week, where I was—of course—also engaged in writing the Great American Novel.

Native New Yorker Carl has a rather more suspicious nature than native Californians such as myself. He had to be talked into even running an ad that would let a stranger enter our lives. I just tried to have faith that the person we needed would find us, using what he calls my "irrational side."

The interview was consequently both intense and thorough, but Jeri was so straightforward and charming that Carl was immediately won over.

He became convinced that he could trust her, and it didn't hurt that she hailed from a small town in Wisconsin, a fact that seemed to him altogether wholesome.

Anyway, we both took an immediate liking to Jeri. Carl explained her duties pretty thoroughly, so she must have liked us, too, because she agreed to take the job.

It couldn't have been for the money, as there was none promised—just a rent-free house on 120 acres in the middle of nowhere with large pets and lots of responsibility.

As it turned out, remoteness was just what she craved.

Divorced for several years, Jeri had lost her youngest daughter—a college freshman—to a drunk driver the previous spring, and was looking for someplace quiet to hole-up and heal.

And so we shook hands and planned to meet in Garberville the following week for the great unveiling.

\* \* \*

Fortunately, as soon as she saw Twin Creeks Ranch, Jeri was so delighted she forgot the arduous and daunting trip.

Jeri was a godsend, freeing our minds for the rest of our busy schedule and allowing us to stay home more and finally get something done on the ranch. She fed the cows and cats, counted noses, weeded and watered and pruned the garden and orchard, and kept the house clean and neat.

Back in Sonoma County, Carl was involved in winding up his property business of 25 years and trying to extricate himself from being a landlord, and I was still working part-time as an interior designer while trying to launch my writing career.

Whenever we'd show up at the ranch, Jeri would have the house not only clean, but warm. We'd visit, of course, and if we got there before dark, we would wander the place, at last able to enjoy it.

We began, somewhat guiltily, to relax. No longer did I have to spend the entire morning after our arrival dusting, vacuuming and reclaiming the house from spiders, mice, wasps, raccoons and itinerant bats.

What Jeri didn't know about repairs, she made up for in welcoming sweetness. The cattle were fed and happy, and our guilt over leaving them alone when we traveled was finally gone.

She had come to us on a provisional basis, for three months. If she was unhappy, or we were unhappy, we agreed to dissolve our agreement at any time that was mutually convenient.

Our collaboration lasted not three months, but three wonderful years—long enough for us to get used to the idea we would have to sell the place and leave one day soon, and long enough for her to begin a new, more positive phase of her life.

And long enough for the three of us to form a lasting friendship that continues to the present day.

It turned out to be just enough.

\* \* \*

While she was in residence at the ranch, Jeri wrote a book, *Backdrop for Joy*, about her experiences of grief and healing. In it, she recalls something Carl wrote and hung on the office wall:

To those who seek
Healing
In this place
There is nothing
You need do —
Relax…
Let Twin Creeks
Work its
Magic
On you.

\* \* \*

Of course, on a ranch there is always work to occupy a mind raw from too much feeling, but there are also natural rhythms to be observed, and the birth of a calf or the opening of a flower can begin the closing of profound wounds.

In *Backdrop for Joy*, Jeri says:

"This kind of life with the physical work involved also requires the balance of rest, relaxation, leisure, stillness…Like a weaving—if all the threads are too tight, the cloth is coarse and rough, stiff and non-workable. On the other hand, if all the threads are too loose, it is sloppy, without form, non-useable, non-functional. There needs to be a balance in the warp and woof, and in the tension."

Jeri knew even in the midst of her pain that there is no way around grief—only through—and that she was able to use our ranch for that sacred purpose was enough for us.

\* \* \* \* \*

## Chapter 22

# Country Roads

*"When you come to a fork in the road, take it."*
—Yogi Berra

WE HAD CERTAINLY TAKEN THE FORK IN THE ROAD; at least we had done it, figuratively.

We had also done it in reality.

Our route was frequently deserted. Sometimes no more than two or three cars passed over that last stretch of road to Zenia in a day.

Whenever I traveled that treacherous 30 miles from Garberville to Zenia, I was reminded of an old song from the '40s that I learned when I was a child: *"Look Down That Lonesome Road."*

On one lonely Zenia trip, I appended my own subtitle: "…and always be sure to wave."

Not just any sort of wave would do, however, there was a very strict protocol.

The prescribed gesture of recognition was a subtle lift of the index finger of either hand, just far enough off the wheel to be noticeable by someone looking for it (a native of the region), and imperceptible to someone who is not. If the oncoming driver happened to be a native, he or she would respond in kind.

Driving that road from Garberville to Zenia was always an experience in being present, for to daydream is to find oneself headed for a very steep drop with no one around to pick up the pieces, or worse, being labeled in one's own mind as a killer for flattening some poor creature on the road.

Road hazards took many forms: rain, fog, snow and ice, horses, cows and donkeys, squirrels, deer, raccoons, foxes and skunks, and, of course, frogs.

\* \* \*

As I mentioned, we often traveled to the ranch in Moby Dick. This was Carl's name for the hulking old GMC truck (a former PG&E workhorse covered with white house paint that had peeled badly, revealing the ugly brown color underneath).

On one particular trip home to the ranch in Moby Dick, I was startled out of a daydream by Carl's rapid application of the brakes.

"What the hell?" I said as I was stopped by the old seatbelt mere inches from the windshield.

He threw the ancient truck into neutral and stood on the parking brake as he pried open the creaky driver's door.

"Wait right here," he said. "I've got a present for you!"

"Wonderful," I said. "Warn me next time."

He bent down in front of Moby Dick's mossy hood and disappeared from sight for a few seconds. When he reappeared, he was carrying something greenish-brown in both hands.

"Here," he said, handing something squirmy to me.

"Oh, great," I said. "A frog. How thoughtful. What the hell are we gonna do with it?"

"It's not a frog. It's a toad. You said you wanted one for your garden just the other day, so, there you are!"

"Oh, great," I said. "You expect me to carry this all the way to the ranch? It'll pee on my hand before we get there. I'm sure of it."

"No, it won't," Carl said. "Put something under it if you're worried."

"Oh, for Pete's sake," I said, removing one hand from under the creature and groping in my purse for a tissue. "Stop wiggling, you," I said, addressing the fidgety reptile as I gingerly shoved the tissue under it, and not a minute too soon, for the toad, as predicted, rewarded my efforts by urinating all over my hand.

"Oh, for God's sake," I said. "It did it! It peed!"

"Oops!" gasped Carl, laughing. "You made him nervous."

"Yeah," I said, as I juggled my purse, more tissues, and the toad. "That makes two of us."

"C'mon," said Carl, when he recovered his breath. "He'll be so happy in your garden eating bugs. And you can make him a really nice toad house.

"I thought you'd be happy. "

"I'm ecstatic. Really."

\* \* \*

One of my mountain lady friends once told a harrowing tale about having a flat tire on our road. This was before the era of cell phones, so she had no means of communicating with the outside world.

Being a thoroughly modern woman, she managed to get the flat tire's replacement out of her trunk. This took at least half an hour, as she grappled with rusted lug nuts and broke a couple of fingernails in the process.

Once she got the tire out of the trunk and onto the ground, she was unsure what to do next.

As she was contemplating her problem, a very old black Honda with clear plastic and duct tape over the rear windows rolled slowly up behind her and stopped. Three very scruffy looking young men hopped out.

My friend was terrified: of carjacking, of robbery, of rape.

But in this case, her judgment was totally wrong. The scruffiest of the three approached her, smiled through tobacco-stained teeth, and asked if she had a jack.

"N-no," she managed to say. "I thought it was in here somewhere, but it doesn't seem to be anywhere."

"No problem, lady," said scruffy individual number two. "We got a jack in our trunk. Tiny," he yelled. "Get the jack," and to her, "We'll have that tire fixed for ya in a jiffy."

*Oh, God, she prayed. Just get me out of this alive and I'll go to church again. I promise!*

Tiny emerged. He was, thankfully, tiny. He didn't look too threatening.

"Jack her up, there, Tiny. I'll get the spare," said the man, who was now beginning to look a lot less scruffy to her.

"We'll get 'er fixed in a jiffy," he repeated.

And they did.

Afterwards, they piled into their car again and took off, waving and smiling.

"Thank you, thank you," said my friend weakly.

"No problem!" yelled scruffy number three, jauntily and smiling. And they were gone in a cloud of grit, dust and flying gravel.

And though she never made it back to church, she did pray a little—at least every time she drove that road. And she always carried a jack.

\* \* \* \* \*

## CHAPTER 23

# The Teacher

*"The decent docent doesn't doze;*
*He teaches standing on his toes."*
—David McCord, *What Cheer*

I HADN'T BEEN AT THE RANCH LONG WHEN I REALIZED I was going to need a source of income, though food was not a problem with an abundant vegetable garden and a freezer full of beef.

Still, I needed gas money and the occasional bottle of sherry, and we had a hefty phone bill to pay (everything being long distance from Zenia), so I surveyed my marketable skills, and interior design not being exactly in demand on the range, I searched farther back in time and came up with an obvious winner: teaching.

In a former life, I had given 15 years and untold amounts of effort to what was at the time a pretty thankless job: trying to teach high school students the value of a liberal education.

As their most frequent request then was, "Can we have a free day?" (whatever that meant), I quickly realized I was going to be swimming solidly upstream; still I stuck it out for that interminable stretch because I was either dedicated or lazy, or a combination of both.

I had taught high school art (too often a dumping ground for washouts from the shop classes and so-called "mainstreamed" special education students) and English (next to math, the most unpopular course in that high school's curriculum). The boys either wanted to be rock stars or pro football players, and thus considered a high school diploma somewhat redundant; and the girls wanted nothing more than to learn how to make meatloaf and sit adoringly at the feet of the boys until they were old enough to get married (16 was universally thought to be just the right age for marriage and family, not necessarily in that order, so obviously high school hadn't much to offer them, either—except, perhaps, a recipe for meatloaf).

I'll never forget the statement of one of the most popular cheerleaders ever to end up in my freshman English class, when confronted with an essay test: "But, I hate these. Why can't we just have a multiple choice test," she said, whining indignantly. "You expect us to think!"

\* \* \*

And so it was that I began the interminable paperwork required to become a substitute teacher in the Kettenpom-Zenia school.

It started innocently enough, with me volunteering to teach the art docent program (a so-called "enrichment" experience).

I had my art history background to call upon, and a seemingly unlimited supply of formerly unused posters of artists' works, from Praxiteles to Picasso, but explaining such complexities as art and artists to children—most of whom had not yet traveled off their mountain—stretched my creative powers to the point of breaking.

So what did I do? I undertook to teach reading, writing, math and science, K-8, with no prior experience.

After all, I did possess a General Secondary Life Credential (one of the last to be issued in the state of California), so I was technically qualified to teach all of those grades. How hard could it be? Besides, I was needed.

At that time, there were only two teachers at the school, which consisted of two adjoining classrooms, separated only by one of those brown Naugahyde plastic folding screens that some idiot once thought would keep out noise and wandering children, but that did neither.

With such a small crew (one of whom was also the principal), when a teacher was ill or had to be absent for whatever reason, there arose a great gulf that had to be filled by calling in an aide to pinch-hit, while the remaining credentialed teacher spent the day ping-ponging between the two rooms in a vain attempt to supervise both classrooms, thus fulfilling at least the letter of the law.

Kindergarten through 4th grade were taught in one room, while 5th through 8th grades occupied the other, so when I subbed, I sometimes got the little kids (sweet, but hyperactive), and sometimes I got the big ones—and now and then they were *quite* big, having been held back to repeat the 8th grade once or twice. These were not so sweet.

In the middle of one of my undoubtedly boring lectures to these older kids, one of the 6th grade girls asked me what kind of rifle I owned.

I made the mistake of telling her I didn't own one, which resulted in an immediate and noticeable drop in her esteem for me.

"My dad's gonna take me out huntin' this weekend," she said. "I got a brand new rifle for my twelfth birthday! You mean you don't even own a gun?"

Alas, it was true—at that time I was conspicuously gunless.

But one day, after a week of trying to get a repeat 8th-grader to sit and read his assignment (not a popular idea with him), I began to reconsider my position on guns.

"Listen," I said to him in one final burst of exasperation, "if you don't sit down and shut up, I am going to call your father."

There was an eerie silence then, as the other students turned to watch what would happen next. This kid's reputation for nastiness was apparently well-known by all but me. *Nobody* ever crossed him.

We stared at each other, he and I, for what seemed like five uneasy minutes.

Then he stood up to his full 6' height and said, menace in his voice, "*Where* do you live?"

Just call me Pistol-Packin' Teach.

\* \* \* \* \*

## CHAPTER 24

# Snazzy and the Amazing Pig Hunt

*"A-hunting we will go, a-hunting we will go,*
*Hi-ho the derry'o, a-hunting we will go."*
— Anonymous, *Old Folk Song*

*"Let dogs delight to bark and bite,*
*For God hath made them so;"*
—Isaac Watts, *Divine Songs Against Quarreling and Fighting*

    The life is rich that has a doggie in it, especially if that dog is a Labrador Retriever.
    Such a one was our Snazzy.

"Snazzy" in the (Figurative) Dog House

Well, let me qualify that. Snazzy was what is commonly known as a "pound pooch," which meant that her lineage was not generally known, but she had a Labrador's black, shiny coat, white front paws and black rear ones, and a white diamond on her chest. She had a Lab's boundless enthusiasm. Her build was that of a Lab, and she had a wonderful set of Lab jowls. Her ears were Lab ears, but her broad head was more pit bull than anything.

Still, when people asked us, "Is she a Labrador?" we always replied in the affirmative. It was just easier.

For most of my life I had been, of necessity, dogless. I couldn't even pet a dog without wheezing, sneezing and breaking out in bumps.

Suddenly, miraculously, I discovered in my majority that the allergy was gone. I could finally have a dog.

And what a dog! When I saw the cute little black-and-white puppy with the floppy ears and big feet at the animal shelter, I fell immediately in love. I named her "Miss Lola Snazzola," but Snazzy, or, "The Snaz," as she was known to her friends, trampled both my heart and my calendulas, and I smiled through it all.

I was a dog person at last.

As a puppy, Snazzy was rambunctious, to say the least. She exuded friendliness to dog and man alike, and could frazzle either one very quickly with her exuberance.

Many a hearty dog would flag after 10 minutes of romping with our Snazzy, limp over to its master, tongue hanging out, panting desperately, eyes beseeching to be delivered from this four-legged nightmare.

Once, we left Snazzy with an unsuspecting friend for a few days while we traveled. Her garden and her dog were both showing signs of exhaustion when we returned, and we never heard a peep from her about a return engagement.

When Snazzy grew to adulthood, we thought she had calmed down a little, but she never lost her native exuberance.

She was willful at times, and at first we thought she was merely stupid when she failed to follow our clear directions, but finally we had to admit she was smarter than we were, because she was able to appear stupid when she simply didn't want to mind.

Our veterinarian had a kinder way of putting this foible of hers. "They can be kind of hard of hearing," he said.

In her eyes I saw a zest for life that put my own to shame, and a need for structure that—I hope—kept me honest and consistent and kind.

If only I had known these simple principles when I was a single mother struggling to raise two children alone, the three of us would certainly have been the better for it.

Each day was new, exciting and wonderful for Snazzy. She played hard, she worked hard, dug vigorously, barked joyfully, slept soundly. She was a happy dog.

On the ranch, she was not just happy, she was ecstatic; running wildly through the pasture, eating cow excrement, sniffing out mice in the barn, digging up hapless gophers and flinging them into the air joyfully before delivering the coup de grâce, wading in the streams, swimming in the lake, mucking through the swamp and barking at deer.

* * *

There were many memorable adventures for Snazzy while at the ranch, but the most memorable was The Amazing Pig Hunt.

Here is how it all began:

One fine morning, Carl had gone down to inspect the lower pasture, where he was struggling to encourage the good pasture grass that grew in the rolling fields below the lake.

What he found was disaster.

Feral pigs had invaded the field and torn the whole thing to pieces. Big clumps of grass lay everywhere, their delicate roots exposed to the sun, dying. Vowing retribution, he stomped back up the hill to the house, seething mad.

"I'll kill the bastards with my bare hands," he swore. The next day, after some deliberation on the best way to go about this task, he called our neighbor, Albert Walpole.

"Sure," said Albert, "I'd be happy to kill them pigs for you. When ya wanna do it?"

"When can you come?" said Carl.

"How 'bout tomorrow?"

"Fine with me," said Carl. "The sooner the better. The bastards are destroying my pasture."

And so it was decided.

* * *

I don't recall that any time was ever discussed, but nonetheless, as I climbed out of bed naked and shivering the next morning at 8 o'clock, I was startled to see the old man seated on his horse just outside our bedroom window.

"Holy shit," I whispered to Carl. "You better get your pants on and get down there, because Albert's here already: he's got his dogs with him, and he's armed!"

"Holy shit," Carl echoed, jumping out of bed and groping for his jeans. "I had no idea he'd be here this early. He must really want those pigs."

"Shall I make breakfast? Coffee?" I yelled at his receding back.

"No time!" he yelled back. I looked out the bedroom window just in time to see Snazzy bound out the door, ecstatically wagging her tail. She was certain they had come just to visit her, so, of course, she did her "let's play" bow.

Now, Albert's dogs were hunters, and that was all they did. Carl had asked earlier if it was all right if Snazzy came along on the hunt.

"I'd like to see if my dog can learn something from the professionals," he said.

"Sure, bring her along," Albert said magnanimously.

By the time I got dressed and made it down the stairs and into the frigid kitchen, the hunting party was disappearing rapidly downhill, rifles bristling—Carl, our neighbor Clancy and Clancy's 10-year-old daughter on foot, and Albert in the saddle of his mare, his rifle holstered at his side and his Stetson firmly seated on his 75-year-old head, but still sitting tall and proud.

Albert's four "pig dogs" swarmed around the horse's hooves, their minds on one thing: wild pigs.

For her part, Snazzy swarmed around the dogs, still trying to get them to play, but they would have none of her. They had a job to do and they were ready to do it. They never even glanced at poor Snazzy.

She romped away into the middle of the pack jumping and barking happily. Carl was exceedingly embarrassed by her enthusiastic behavior, but he kept his head high and tried to ignore her.

And that was the last I heard from the hunters until there was a volley of gunfire down near the swamp at the bottom of the property.

It was nearly noon, but I continued with my fire building, breakfast making and housecleaning, until I heard a shout down near the pasture gate.

"Go get the camera," yelled Carl. "We got a big one!"

I raced to find it to memorialize all this unusual activity, but of course the camera had hidden itself again, as it so often did. By the time I got out onto the deck and got the damn thing focused, they were coming through the lower pasture gate and, in the lead, sitting ramrod straight on his horse, a rope tied to the pommel of his saddle, was Albert.

Soon, what was at the other end of the rope came into view. He was dragging a huge black boar through the field, the rope held fast through the animal's jaw.

Snazzy seemed to be the lead dog this time. She was

leaping at the hog, and then leaping back, only to leap brave-
ly on the dead hog again and again.

I snapped away with the camera as the procession
came proudly up the hill. What to me was a humorous scene
had obviously not gone well for the pig.

It was only after the men had said their goodbyes and
Albert had taken off with the huge wild boar that I got the
whole story, and it was a good one, all right.

"Did you get pictures?" Carl asked, coming in the
door winded and stomping to get the mud off his feet.

"You betcha," I said. "So tell me all about it."

"You shoulda seen those dogs of Albert's," he said.
"Old Snazzy kept trying to play with them all the way down,
and they were just focused on pigs.

"And, boy, did they do a great job! We hadn't been
there long before they flushed out a bunch of them down in
the swamp. The dogs went tearing off after the pigs, and we
went tearing off after *them*.

"Snazzy jumped in that swamp after the dogs. There
was a flurry of squealing and barking, and pretty soon she
came flying out again. She'd seen the pigs and she didn't
know what to do, so she kinda hung back behind me.

"Clancy went in after the dogs, and shot the pig," he
said.

"Then I went in, and Clancy and I gutted the pig.
Clancy found the liver and said, 'Maybe your dog would like
the pig's liver,' which I thought was really generous, as the
liver is considered a delicacy.

"But when I offered it to Snazzy, she sniffed it tenta-
tively and then backed away from me, and would have nothing to do with it.

"Just like Saint Peter hanging around Jesus' trial,
Snazzy kept her distance.

"Albert called his dogs, but they were long gone,
baying after the rest of the pigs.

"He said, 'Aww, they'll come on back when they're ready,' so we ran another rope through the pig's jaw and Albert dragged him back with the horse's help.

"I was mortified at my dog's behavior. Snazzy kept well away from the pig as it was being dragged up the hill, but then she got braver and braver as we got closer to the house, so she started making passes at the dead pig, biting its tail first, and then, by the time we got up close to you, she was lunging at it and biting its ears like she was taking credit for the whole thing.

"I guess she finally realized it couldn't do her any harm—after all, the pig wasn't moving too fast at that point.

"She was such a chickenshit I couldn't believe it. That's the last time I take her hunting with me!"

It was an ignominious ending to one of the most exciting times of Snazzy's life, but only in Carl's mind. She took the whole experience with her usual equanimity.

Snazzy went back to barking at deer; Carl gave the boar to Albert, who seemed pleased and shared the meat with Clancy, and Carl and I went back to the many chores we faced daily (none as exciting as a pig hunt!) at Twin Creeks Ranch.

\* \* \* \* \*

## CHAPTER 25

# Hills of Gold

*"All that glisters is not gold—*
*Often you have heard that told."*
—Shakespeare, The Merchant of Venice

*"...rural folk, plain and simple Americans who still live close to*
*the land, who don't care to put on airs, who prefer to remain the*
*way they are... These...are the people of the Edges, those caught*
*on the interface between a traditional country way of life and the*
*forces of modernization that threaten to annihilate it."*
—Ray Raphael, Edges:
*Human Ecology of the Backcountry*

THE HISTORY OF THIS PART OF NORTHERN CALIFORNIA
is rich and varied—both joyful and shameful. The early white
settlers here, like settlers everywhere, saw land they liked and
took it, regardless of others' rights or prior claims.

Much of the early settlement began in Round Valley,
in Mendocino County, and gradually spread north and east
to the mountains. The pioneers began to move farther north,
to ranch and build villages, the names of which evoke their
colorful early life: Poison Camp (now Zenia), Robber's
Roost, Burnt Ranch, Fort Seward, Mad River, Hayfork,
Peanut.

In some parts, the Indian names were retained—
names such as Hoopa, Hyanpom, Sequoia (now Whitlow),
Lassic and Hettenshaw.

Until 1853, Trinity and Humboldt counties were one
very large county, stretching from the Trinity Alps to the

Pacific Coast. In the end, Humboldt got the coast; Trinity got the mountains.

During the 1880s, before the advent of the Northwestern Pacific Railroad, supply boats came up the Eel River, and mule pack trains carried supplies to the widely scattered communities. Each pack train had a "bell mare" which led the train, and two to six men to steer the horses.

There were no roads, and moving to a new homestead meant loading everything you owned on wagons and going as far as you could, then, when the terrain got too rugged, wagons had to be converted into sleds, to be pulled up and down mountains by horse and manpower—whatever did the job. The travelers would lodge for a night or two with whomever they could find along the way (as there were no hotels), and then continue on.

Just as there were no hotels, there were no bridges, either, so streams and rivers had to be forded whenever possible, and that meant never in the winter. There are many sad stories in the early lore about unwary folks getting swept away by wild rivers; the result of foolishly trying to ford too early in the spring.

The pack trains went both ways, of course, so as the ranches became more established, the trains began to carry wool and other local products out with them and back to civilization. It was in this way, too, that the area's first families received their mail.

This delivery was also dependent on the weather. If it was snowing or raining too hard, or rivers and streams could not be safely forded, there was no mail, period.

Lumbering was a major industry. Until the 1980s, it was the largest industry in Trinity County.

Most of the county was originally covered by lush green forests, which served—and still do—as watersheds for the Trinity, Mad, Eel and Van Duzen rivers. Development has made few incursions there, but the watersheds and the

rivers are now as hotly contested between loggers and con-
servationists as are the trees themselves.

In recent history, there were as many as seven large
mills in the county, each averaging about 100,000 board feet
of lumber per day.

By the time we arrived in 1990, the lumber industry
was all but dead, a victim of the environmental movement
and the Law of Unintended Consequences.

* * *

Despite the hardships inherent in living in the
wilderness, most who came adjusted well to the new life—
hunting, fishing and trapping to supplement the larder. The
women raised chickens, turkeys and hogs, and planted vegetables and flowers.

Life was tough, but it was real, and often a man could
be his own boss, answering to nobody except his wife and his
God; even living off the land if he was resourceful, and, in
general, being content.

Some of the men came for the mining, for there was
manganese, silver and copper in these hills. A big copper
mine at Island Mountain featured a solid mile of tunnel. The
ore crossed the Eel by cable and traveled all the way to
Tacoma, Washington to be smelted.

It was said that enough gold and silver was recovered
during the smelting process to pay for smelting the copper.

But it was the lure of gold that brought the first settlers to Trinity County. People panned, picked and shoveled
for it in the creeks and rivers, and when that was gone, built
giant placer mines and dredges. They dug tunnels and blast-
ed with dynamite. At one time, the La Grange Mine near
Junction City was the largest hydraulic mine in the world.

Today you can still see a mining claim site or two,
and the still-naked mounds that once were lush mountains,

their flora and topsoil having floated down the rivers through hydraulic mining, clogging them for miles and burying cattle and barns in the valleys below.

The lonely road over Grizzly Mountain is still marked here and there by a hand-lettered claim sign, nailed to a tree.

The perennially hopeful still roam those hills, gold pan in hand. With so many folks on welfare (having been laid off from logging jobs during the economic slump of the '80s and '90s and the relentless incursions of radical activists, with deceptively sweet names such as Verbena and Morning Glory), there was not a whole lot else to do.

\* \* \*

In the old days, many of the new folks raised turkeys for market. These had to be driven down 90 miles of primitive trail, all the way to Eureka to be sold.

The November 18, 1911 *Blue Lake Advocate* stated:

> Mr. and Mrs. Abe Bush and three children of Hyanpom Valley, Trinity County, arrived here Monday after a five days trip. With the assistance of a good dog, they drove a band of 112 turkeys, which is no small task considering the steep and narrow trails, over Grousse Creek Mountain. They only lost a couple on the way. The turkeys were bought by the Northern Redwood Lumber Company of Korbel for $2.35 apiece, which will furnish a first-class Thanksgiving dinner to the company's many employees.

\* \* \*

At first, sheep-raising seemed the way to go if you had some land, but wolf and coyote depredation proved too severe. Gradually, the new settlers began to raise cattle, and unemployed men often found work as cowboys.

Many of the cattle came from wild herds that had wandered off from the ranch of the wealthy Indian agent in Round Valley—cattle that were originally intended for the Indians, but that somehow (like much that was promised the Indians over the centuries) never got to their intended destination.

The first white-faced Herefords (the breed we raised, and, according to the Trinity County Cattlemen's Association, "the most popular breed of cattle in Trinity County") were brought to Trinity County from Sonoma in 1895.

The ranchers in those days had to contend with wild hogs, which ate dropped calves. The boys on the Merritt Ranch solved this problem by making hams and bacon to sell in their father's general store.

When we were there, cowboys still drove cattle along the dusty dirt roads to pasture, but not to market—that trip was made in style in big cattle trailers, and avoided the animals' inevitable loss of weight when the only way to get to market was to walk.

But the modern herds still needed to be rounded up and driven out of the Trinity and Six Rivers national forest lands (where they generally grazed throughout the winter and spring), and back to their home range for the dry summer and fall seasons. According to a recent article in the *Red Bluff Daily News*, "Currently in the State of California there are four counties that have been designated as grazing counties, Shasta, Siskiyou, Trinity and Lassen counties."

As Trinity County is mostly national forest (75% of the county's land is administered by the U.S. Forest Service and Bureau of Land Management), there's plenty of room for cattle to graze, and much of the county land is still open range. Cattlemen pay a fee to the Forest Service for the privilege of grazing their cattle on government land.

Without this agreement, which since 2003 has been limited to 10 years, it will become increasingly difficult to raise cattle in Trinity County.

It has always been a perilous living—and cattle rustling still exists, though the penalties are not now as steep as they once were. Exhibit A, in the words of one Southern Trinity pioneer woman:

> A few years before we came here, the Littlefield hanging took place near the Double Gates and French's Lake, (near Kettenpom) and many of the settlers who wanted law-and-order took this way to stop the rustling of cattle that was prevalent here. How right they were is not for me to say, but none of that wild life is here at the present time.

She continues with an interesting tale of how cattle were driven to market in her time:

> In those early days cattle sold for three or four cents (per pound) on foot...The cattle were wilder than the deer and many a steer was chased for days before being caught and tied up. Dogs were trained at that time to track cattle.
>
> The older settlers had trained oxen to lead out the wildest steers. They were hitched together like a team of horses...and the trained animal led out the wild one(s). When they started on the long drives taking the cattle away, they would sew up the eyes of the wildest ones with buckskin string and whenever one left a bunch, he was caughtand his eyes (eyelids) were sewn up. They very seldom lost or had any cattle get away on these drives.

* * *

After 1914, when the Golden Spike connecting Eureka and Willits was driven two miles south of Alderpoint, life became a little easier on the mountain. Many men were employed by and for the railroad during the building, and farmers made money selling fresh fruit and vegetables to the railroad gangs.

Supplies could then come by railroad freight wagon from Fortuna, 80 miles away. You could also order your clothes and work boots from the Sears or Montgomery Ward catalogs and have them delivered by parcel post (only problem was, the weight limit on parcel post was four pounds, so a pair of work boots had to be delivered in two packages).

When, in 1920, telephone service came to the area, it also meant greater contact with the outside world, but that contact remained limited even up to our time, because very few families in the area had telephones—it just cost too much, as one had to pay for the stringing of telephone lines by the foot.

The surest means of communication in 1990 was still a remark to the postmistress or a note left at the general store.

The social scene on the mountain was lively in the old days. In the early years, before the dance hall (now long gone) came to Zenia, many of the early dances were held in the old schoolhouse. Everybody brought sandwiches and cake for the traditional midnight supper, where coffee was served from a five-gallon can.

At these dances, they typically had two musicians: violin and guitar, plus a caller for the Quadrilles.

There was no school held in winter, as the weather was too unpredictable, but if dances took place during that season, the local women often got stuck with out-of-town house guests for longer than they'd planned.

The local folks were also fond of having big picnics, which featured horseracing and steer roping. Fourth of July celebrations lasted for days. Picnics at Blocksburg drew crowds of up to 2,000—mostly Indians—from the surrounding countryside.

The Grange Hall (Community Hall) was built in 1915. Two hundred people attended the dedication celebration. Lumber was hauled long distances over new wagon roads for much of the construction.

Gradually, the quality of life on the mountain improved, but the living of it became no less precious. In the poetic words of one of Trinity County's early women settlers as she reflected on her life at the age of 85,

> I am a native daughter of California, and I have never traveled far or wide, but I have seen the deep purple shadows over our mountains in the autumn.
>
> I have seen the Alpine glow on the snow-capped peaks of the South Fork Range at the close of day, as the sun made its final dip into the distant sea.
>
> I have seen a silver thaw at midnight under a full moon; the most exquisite fairyland one could imagine, with every twig and branch a sparkling diamond.
>
> I have seen a rainbow at midnight over the Kekawaka Canyon, and I have seen our great trees bow their heads under their weight of snow in silent tribute to their heavy burdens, then raise their leafy arms to Heaven in praise, as those burdens slipped to the ground.
>
> I have listened to a great white silence so intense that one senses the presence of God alone.

I have listened to the wild storms that roared as they swept overhead, and the gentle lullaby of the winds through the pines as they lulled us to rest.

I have held in the hollow of my hand a tiny hummingbird too weak to follow its family to their winter home—the most exquisite work of art I have ever beheld, each tiny feather blended into the other, and covered with an iridescent glow...

I have gathered armsfull of the fragrant Redwood lilies. "They toil not, neither do they spin," and yet they are so beautiful.

I have seen the pine drops the color of deep wine; buds, stems and all.

I have seen the ghost flower with its long spikes of tiny white bells, and the lovely lady's slipper, deep within the forest.

This is truly a man's country, but...the women made a great change...life changed for me as a pioneer woman, and with a patient hand I tried to remove the briars in my path.

How well did I succeed?

I am here to stay, and I learned to live the wild, rugged life at its best on a wild, rugged mountainside in Trinity County.

And, like Ruth, I said, "Thy people are my people, and thy home is my home."

—Jessie E. Gummer

\* \* \* \* \*

## CHAPTER 26

# Cowgirl Alone

*"Wholeness either of body or of psyche can never be achieved by man deciding from day to day whether to work for it or not...unless he is dedicated to it now and in the future—it is not likely to happen."*
—Carl Jung

*"To live authentically in the present moment, to meet challenges head-on; to be truthful, not fearful—these are things I must always endeavor to do."*
—ML's Journal, 1996

*"All men should strive to learn before they die
What they are running from, and to, and why."*
—James Thurber

IN 1996 I MADE A DECISION THAT I NEVER EXPECTED to make. I felt a strong need and desire to live alone at the ranch. Something in the place itself called to me to do it.

It was uncharacteristic of me. Not that I minded being alone—as an only child until the age of 15, I had learned to be alone and loved it—but to be alone in such a remote place—that was new.

Nevertheless, I found myself gradually in possession of a pressing need to be absent from the world and just *be*. Not to drop out so much, as to check in with myself, perform a few course-corrections, evaluate where, if anywhere, I was going.

Carl was fine with that because, just then, he could only be at the ranch part-time and it relieved his mind to

*187*

Cowgirl and Cow—Alone

have me watching the place and feeding the cattle. And then, too, we still had cats. The ranch cats needed at least occasional attention and food in addition to their diet of birds and mice and gophers and moles.

I found that I loved it, this being alone, really alone.

Perhaps the word love is too weak for the joy I felt as I woke each morning true to no timetable but my own inner clock and responding only to the rhythms of daylight, sunset, heat and coolness as I planned each day.

Yet each night I had to face my fear.

I had so many fears then; fears I hadn't known I possessed. In the more violent world in which I now lived, I was fearful of trespassers, of teenage kids with too much time on their hands who lurked, I was sure, just beyond the safe boundaries of my locked front gate.

I was fearful, too, of the one thing that should have made me feel safe: the antique .32 Smith and Wesson revolver that had belonged to my grandmother, the one my mother had given me when I moved to the ranch.

I'm terrified of guns in general, and especially handguns, since, in April 1971, I watched, horrified, as my husband of 14 months used one to shoot himself through the head. He fell dead at my feet, spattering my ironing basket with his blood.

Learning to shoot the rifle under Carl's expert tutelage was only half as frightening as holding that small, rather feminine gun in my hand.

I feared injury and the subsequent agony of trying to get help from wherever I was on the land, alone and perhaps without the dog for company, as I had very little confidence she would stay with me or go for help.

Knowing her, I figured she'd just go off chasing a deer or wander away because she was bored—she had never shown a high degree of loyalty or bonding to either of us.

I feared harming another human being.

But my greatest fear was that archetypal, primordial fear of the dark side of Nature herself.

Nature not in the abstract, but up close and real—untamed and unmanageable and little understood by those of us who had always lived in the more civilized, circumscribed world of suburbia.

Part of me knew this enforced confrontation with the elemental side of my world was a good thing, representing as it did a largely unacknowledged dark aspect of my own nature—a part of myself and the world which I had simply refused, up to then, to do business with; but part of me also knew (or strongly suspected) that I would be unequal to the tasks presented by such a primitive world, and this part of me keened with raw panic whenever I gave it my attention.

Sure enough, the tests were everywhere: from the dead bluebirds in the woodstove to the dead cows in the creek.

Signs and portents.

Challenges to be faced, small and large.

It started with the chimney pot mystery.

We had at the ranch a large and vocal number of California bluebirds, one of the loveliest and most magical of birds to me, next to crows—which always seem to appear when something wonderful and amazing is about to happen in my life. Crows were, and are, my personal shamans, and they always make me pay attention.

When I first started seeing the bluebirds all over our land, I took it as a sign, of course, that this was the most blessed of all places.

But there was a flip side to this fuzzy wonderment.

After we had been there a year or so, I would come back to the ranch after a few days' absence to find a bluebird had somehow flown past the chimney screen and dropped dead in the Franklin stove we used to heat the house.

After the first two dead bluebirds, I developed a real reluctance toward opening the stove to build a fire (one of my favorite things to do), fearing yet another feathered kamikaze would be lying there, feet up and stone cold dead in the fluffy ashes of the Franklin stove.

The bluebird of happiness. My good omen bird—its life extinguished. It was just too much.

And then the cows started dying, as cows do, as a result of their nature, which I had also revered perhaps excessively.

\* \* \*

The first to go was Limpy, a cow we inherited from the Scotts when we bought the place, and barely got to know before she exited this vale of tears.

She died in childbirth—another fear of mine, but my own two labors had been not merely lengthy but also joyously productive.

Cattle cannot always handle their lying-in by themselves.

It was a breech birth. Limpy had gone off to be alone, to find a little dignified privacy in which to do her most important work, and she had died wedged between stumps and rocks in little Alder Creek, not far from the bosom of the herd; but worlds away for her.

Our itinerant cowboy, Jimmy, found her there when he went to check on her, only a few days after we had left the ranch to take care of some business in town.

We grieved, and hoped without naming it, that this was not the ranch's way of telling us we were not welcome.

I did what I do best in such circumstances: I simply didn't deal with it, avoiding the whole event until there was nothing more to see at the death site but a lot of unrelated bones, scattered by circling vultures and coyotes, until Limpy no longer resembled the skeleton of a very large cow—*our* very large cow.

Next came the sacrifice of Hamburger, our inherited steer. He was a silly brown, white-faced clown, and all our efforts to stay unattached to him failed, as we learned for sure when we said goodbye to him for the last time at the slaughterhouse.

With hearts more heavy than we could have imagined, we abandoned him to his fate—one which we had consciously imposed upon him—and drove home crying, feeling like murderers, to write poetry in his honor and try to expiate our feelings of guilt.

* * *

Next, there was Cisco, a steer born of Big Mama, to be raised from a nursing calf solely by our inexperienced hands.

Cisco was a black-and-white, spotty faced babe born in the woods with the herd, and Carl, attending.

He was a sweet baby, but soon developed less endearing traits. We discovered why on close examination of his business end.

Cisco had been one of our first attempts at castration by the plastic ring method, and, as it turned out, we were not entirely successful.

We had managed to trap one but not both of his testicles within the plastic ring, where it had atrophied, as it was meant to do. But the other was still there, large and resplendent and capable of turning our docile baby boy into a fury of a bellowing, howling bull.

Of course, he had to go. But first, I had to feed him twice daily, fattening him on grain for the eventual slaughter. I felt conflicted about this task, of course, and when Carl took him down the mountain for that last ride, I was sad.

I cried when we said goodbye. At least I was saying goodbye. My macho man declared that as he had seen the creature into the world he would see him out, and planned to drive back to the meat man's pasture in the morning to witness the shooting.

Unfortunately, our half-bull, half-steer had challenged the butcher's resident bull, and the resultant bellowing kept them awake all night, sealing an early fate for Cisco.

Carl got there in time at least to observe poor Cisco's dismemberment.

Then there were a couple of cute steer calves that went to the cattle auction in Fortuna. That was easy, as we could only guess their eventual fate, consoling ourselves with fantasies of wholesome 4-H children who would purchase our steers and give them a least a year of pampering before the final sacrifice—sort of like the Aztecs and their victims.

More female problems ensued, with our sweet Baby developing a prolapsed uterus. Several attempts to reinsert the inverted thing failed, so she had to be put down, and by Carl, as she was suffering.

Our herd—despite the deaths—increased nearly exponentially, until at the last we had 26 head of cattle, each one, against our better judgment, named by us, and each one exhibiting a distinct, unique personality.

Pansy was a wild and feisty one—black with a white face, which had three large blotches of black upon it in a pattern that made her look just like her namesake flower.

Alas, her personality was not at all flowerlike.

Pansy was pushy and impatient, and always arrived first among the heifers when it was time for alfalfa. I must confess that, despite my naming of her after a favorite flower, I didn't much like her. And so it was that the universe handed me yet another opportunity to deal with my fear and guilt.

Pansy had already had one calf successfully, so we didn't expect any problems—indeed did not even think she was near to giving birth—but she was smaller than our other cows, and it was hard to tell.

But, returning from the city alone after a business appointment, I found Pansy, too, had died as a result of another breech birth. I found her in Alder Creek, wedged belly-up between some boulders, her head thrown back, neck arched in her final agony, a pair of small hooves protruding from the birth canal.

Alder Creek and Rocks in Dappled Shade

So sad; so unutterably sad. When I found her, she was already stiff, her hooves in the air, bent at the hocks, and the flies had begun their inexorable work on her snowy white belly.

The vultures waited.

I forced myself to look carefully at her so I could be sure what had happened, circling ever closer, like the circling carrion birds, to that which I was trying so hard to deny. The messiness of death.

The difficulty of including that fact in my world.

I was in the habit of casting a mental, protective ring around the property when I arrived, sort of a charmed circle within which nothing bad should ever happen.

Yet I had been forced to factor death into my safe and private world.

If anyone had told me when I met the distinguished, urbane, well-educated and well-traveled man who is my husband, that I would soon be up to my armpits in real life, wading through cow shit, dealing with large, dead animals, and becoming a reluctant expert on rural plumbing systems, I'd have told them they were seriously deranged.

After all, I was not about to trade in my silk designer suits and lavish client lunches at the Galleria in San Francisco for overalls and cowboy boots, even if the latter *had* become trendy in my fashion-conscious circle. No, my idea of country life consisted vaguely of wearing lots of tweed with leather patches, stylishly paired with desisgner jeans (the $250 variety), and maybe a hearty plaid flannel shirt or two, so I could look the part of the wife of a proper country squire as I shopped for delicacies in town with which to entertain the frequent spate of city refugees we were sure to attract to our fabulous mountain estate.

I would host picnics on the closely cropped greensward with checkered and calico tablecloths in bright colors, flowers from my own weed-free garden, dishes featuring homegrown veggies lovingly prepared by yours truly and flavored with a huge variety of all the herbs I grew with my own uncallused hands.

I'm quick, though. It didn't take me more than a few months of backbreaking labor to figure out that most of the folks who achieve this kind of ambiance (Martha Stewart included, curse her) have a staff of 25 or so behind the scenes to help them pull it off.

That made me feel a whole lot better.

* * *

But there was more going on than the shock and sadness I felt when I thought about all the deaths going on around me on the place we called Twin Creeks Ranch, more than the finding of smashed birds' eggs, abandoned nests, the skulls of small animals, and the carcasses of deer brought down by men and beasts.

I had to face my fear once and for all.

It was a paradox I had never come to terms with, but at least I learned more about myself, and my fear of being alone. The paradoxical part was, that although I loved being alone, I also hated it, because I missed Carl so terribly.

I didn't miss his messiness, as the house was much easier to keep up when he was not there, and I didn't miss his looking over my shoulder as I cooked, gradually taking each dish in a different direction than the one which I had intended. I must admit it was very, very quiet and peaceful at the ranch without such robust male energy around to stir things up.

What I missed was his presence in the quiet moments of our lives, sharing our awe in beholding the mysteries of Nature as they unfolded before us in that most wondrous place.

I missed the physical closeness of him, the smell of him, his walk, his infrequent, blessed smiles.

\* \* \*

Of course there was a practical side to my loneliness as well. It was just more reassuring to know there was someone else there to figure out how to fix things, to take care of the many lives that were our responsibility, to help hold back the fear of being alone.

Someone who could shoot straight. Someone who cooked.

I think still the most frightening thing about fear of being alone, though, is the fear of facing oneself. When one is alone, the internal monologue that goes on ceaselessly in the background of each life must be heard.

And sometimes, I'll admit I just didn't want to listen to what it had to say:

- That I was frittering away precious time, reading and napping in the hot afternoons;
- That I had no writing talent and was just faking it— pretending to be a writer;
- That I didn't deserve to be happy in such a beautiful place without sharing it;
- That I sometimes had great difficulty getting down to the business of writing;
- That I had let the cows down by not being present in their very hours of need;
- That this enterprise we had embarked upon was sheer folly after all, and our accomplishments here would never be enough for us to be worthy stewards of this beautiful land, which had been so lovingly cared for by the Scotts before us;
- That I was not taking my spiritual life as seriously as I should;
That I was blowing it.

Nobody wants to hear those things, of course; especially not when they're self-imposed, yet the voice in my head was finally heard in the silence of the country, where it could not be denied. *Take a risk*, it said. *Only then will you know your true self.*

And finally I heeded what it had to say.

\* \* \* \* \*

## CHAPTER 27

# Letter From Zenia

April 4, 1999 Easter Sunday

It is a rugged place this, inhabited by rugged people. Its purple mountains' majesty is undermined by human pain, and there is plenty of that to go around here.

Here live unemployed loggers, part-time waitresses, retired schoolteachers, starving artists, struggling ranchers.

Here too, no doubt, live wife-beaters, potheads, petty thieves, drunkards and hard-working, decent, honest men.

Here live nagging wives and midwives. Shrews and healers. Here live patient, strong and saintly women.

I live here as well. Not sure of a category where I fit, though I'd like to think I do.

Sometimes from here I cannot see the shining ocean through the landscapes downrange, which tumble, eventually, into the sea; into the ether beyond us all.

Sunrises are different here, filtered through a fringe of tall firs, but sunsets are magnificently unfettered, as the sun melts into the lap of the mountains.

Blessed Trinity.

Birds twitter; silence surrounds all. Mornings are holy, pristine, crystalline, advancing in a just progression out of velvety night.

Each day is a rebirth, and spring makes long winter worthwhile. Spring is a delightful bursting-forth of new life in meadow and forest: a cellular energy which builds in secret through the dark months until it can contain itself no longer and ripples out into the world—cosmic dust shaken from God's great cloak, which settles on the earth as wildflowers, blooming grasses, fruiting trees, baby deer and brush rabbits and bluebirds, baby cougars, coyotes and rattlesnakes, bobcats and bears.

And for us, baby cows.

It's like nothing I've ever experienced.

It's wonderful, terrifying, fulfilling and frustrating, liberating and confining.

It is my greatest joy.

\* \* \*

April 13th, 1999

Lengthening shadows. I sit on the deck and enjoy the breeze on my sweaty skin, and the myriad of sounds: the sound of a frog croaking, a stiff breeze, so that the wind-direction indicator flaps back-and-forth, creaking softly; the wind chimes whispering.

All is underscored by the plangent sound of the creek—Alder Creek—which flows and tumbles from its source here in the mountains toward the sinuous green Eel River below.

That, and the silence.

In fall, the Eel River really looks like one— lithe, lean and snaky.

The buckeyes lose their leaves first. Rattlesnakes come out and languidly drape them-

selves on hot rocks. Wild blackberries and rasp-
berries are ripening and wild carrots dot the
roadside. Everything is dry—waiting.

But now it is spring, and all is lush, green
and wondrous; the way it was when we first saw it
and fell in love.

That first year I went a little crazy with love,
I'll admit.

I planted:

*   Santolina. oregano, rosemary, artemesia,
*   French tarragon, rockrose, salad burnet,
    sage, lamb's ears;
*   Curly tansy, creeping thyme, curly thyme,
    lemon thyme;
*   Curry plant, rue, yarrow, daylilies, lavender,
    mullein;
*   Nandina, penstemon, coreopsis, foxgloves,
    coneflowers;
*   Sweet woodruff, evening primrose, bearded
    iris, coral bells;
*   Feverfew, pyrethrum, money plant, love-
    in-a-mist;
*   Edelweiss and New England asters.
    —And that was just the perennials.

\* \* \*

August 19, 1999

The cows left yesterday, with Arch and his
wife—6 of them, to go over the hill to market.

We're now down to 18, eight of which are
nursing new babies. Had to say goodbye to Latte
and Mocha, Pixie and Latte's baby, and two
unnamed Hereford yearlings.

I took apart and stripped and painted and

reassembled the bureau for our room to keep busy, so I wouldn't think about the absent cows. We also tried to re-castrate a very large male calf—two more to go. Hogtied him on the ground and he kicked and hollered like crazy. Don't try this at home.

\* \* \*

September 15, 1999

Arrived at 3 P.M. I came alone this time.

As I came down Justice Road toward the gate, I flushed a doe and her spotted fawn out of the brush. They ran down the road in front of me and turned off at the gate, disappearing back into the woods. At least the doe did.

The fawn ran the wrong way—toward the gate and me. He froze. Ran back toward mom. Then as I got out of the car to open the gate, he came back for a second look—within five feet of me—looked straight into my eyes, hesitated, and then crashed into the woods.

Small animals' curiosity seems to know no bounds. All I know is, I was as startled as he was!

September 16, Thursday

Smoky all day from fires in the mountains east of us. Gorgeous sunset as a result. Heard coyotes last night, just before dawn. Cows started bellowing just after dawn, so neither Jeri nor I got much sleep, but, as it was her turn to feed the cows, I luxuriated in bed, indulging in my favorite vice—drinking coffee and reading.

Cleaned house and threw out a lot of junk I'll never use. It seems to grow toward me from the corners of a room. Jeri made a Salade Niçoise for supper with fresh tomatoes, feta, fresh lettuce, cucumbers, onions, tuna, olives and oil, plus fresh-from-the-garden basil and savory. Yummy! Not to be outdone, I made my rustic pear tart with walnuts.

* * *

September 29, Wednesday
    Still smoky—fire still burning somewhere. We've been working the past three days to repair the damage done by some hunters on the first weekend of deer season.
    Same thing as last year. They cut our fence to drag through a buck they shot on our land, and the cows wandered down to the neighbor's place below us, eight of them, led by the itinerant bull, and trashed the neighbor's spring and his orchard. He was very angry.
    Called at 7 A.M. this morning and said he is sending us the bill.
    We had to find the cows and call them back, set up a water trough on the upper part of the property, get water going into it, and then lure the cows to the tank. We did this by the following method: we filled the back of the truck with hay, then Carl drove Moby Dick slowly up the hill, while Jeri and I hung off the back, scattering little handfuls of hay like Hansel and Gretel scattering breadcrumbs, as the cows all trudged behind, trying to catch up to us.
    They don't usually like to move that fast, but when there is hay involved, they perk up and pick up the pace. The smart ones ran. Ghost and

Peaches actually licked hay off the tailgate as we went—a traveling manger!

Both springs are still producing, but barely, so we are all on water rationing. Praying for rain. Yesterday, all day was spent on water system diversion and repair.

Today, Carl rewove the fence temporarily and we fooled with the water systems some more. I cut out some recipes from Gourmet magazine and we tried two for supper tonight, using left-over spaghetti and the produce that we and Jeri grew. Good.

Tired.

* * *

November 10, Wednesday

Yesterday I rescued a bluebird from the fireplace where it had fallen, the last of three I've found there. I was so thrilled, to catch and hold that beautiful bird in my gloved hand. I took it onto the deck and launched it into the air, and it flew. I called Carl, who was down at the barn, to see it fly.

It was wonderful.

This morning I saw a flock of crows down by the pasture gate. We were down there to repair the mangers yesterday, and today we cleaned the gutters. It rained all night.

This morning, Woodstock and his wife came around to negotiate a price for the new fencing we need at the bottom of the property. She had coffee; he had beer. The dogs played.

December 31, 1999

Tonight the new millennium dawns. Please, God, may it be a peaceful one for us all.

* * * * *

## CHAPTER 28

# Another Letter from Zenia

August 27, 2002

The KZVFD (Kettenpom-Zenia Volunteer Fire Department) barbecue was quite an affair. Beef and pork ribs, corn, beans, fresh bread and fruit—all for $7.50. Most all the folks were there: Albert and Angie; Jimmy and Jeannie; Joan, of course; fire commissioners, firemen and their wives and sweethearts and assorted community members, new and old.

Lots of kids and a few dogs, although fewer than usual, since dogs have been banned from these gatherings, at least officially.

A liberal sprinkling of local color: Indians and cowboys, women with far too much makeup, and, of course, old Homer Bradshaw, with his leathery face and lean body, cowboy hat and boots, starched shirt and clean jeans. He still looks fit, though he must be pushing 80.

The siren sounded on the new fire truck, and a roaring good time was had by all.

\* \* \*

August 28, 2002

We're heading for the last roundup. One more day here for me—ten to twenty more for Carl—if the escrow closes on time.

Neither of us wants to leave, but the whole ranch enterprise has, with our rapidly advancing age, become just too much for us, though it is a hard thing to admit. We gave it twelve exhausting years, and now every time we drag ourselves up from the garden, barns or pasture to the house, we are reminded that our predecessors were about our present age when they, too, decided that love of this land was not enough.

It is time to go.

Carl has been working nonstop. It shows. He's done wonders with the place (for a guy). Made it look truly lovely, even with a minimum of furniture and accessories. He had flowers on the dining table when I arrived, and my favorite old lace tablecloth. Looked so inviting.

My lavender is blooming, and the lemon balm, rosemary and sage are huge and spreading. (Part of *my* legacy.)

As I write, we're down at Burgess Creek, with doggie muddying up the pristine water. She laps the coolly inviting stuff straight from the little waterfalls that pour noisily over mossy rocks.

Most of our stuff is already packed, so we're just here to have fun. What a concept!

Slept late this morning and breakfasted on the deck, watching the butterflies (California Sisters, Yellow Swallowtails and Hairstreaks) playing in the water spray from the sprinkler on the back lawn. Birds singing. Sun shining. So good!

And so sad.

August 30

Last night—my last night here, we saw a huge brown stag beetle on the front deck, nearly as big as my fist. He clung tenaciously to the deck, even as Carl tried to dislodge him to protect him from being eaten by the dog.

He looked just like the one I carved on a linoleum block back in my college days, but I'd never seen a live one. An amazing creature.

Carl finally pried the beetle loose and dropped him gently to the ground, where he landed with a soft thud.

So many wonderful birds and bunnies and snakes and butterflies I saw on this trip. And deer. We saw several bucks last night on our walk to the Teapot Dome to place our last rock under the little buckeye tree.

It occurred to me around then why deer are so precious to me. It is because they're such symbols of primal innocence. So shy. So vulnerable.

Of course, up close, a stag can be pretty intimidating. We tried never to get too close.

\* \* \*

I photographed everything I could, including the view of road and trees down from our new gate, as it winds so beautifully through the high golden meadow.

Today, I said goodbye, yet not goodbye.

The ranch—Twin Creeks Ranch—Cloud Lake Ranch, will live in memory always.

We looked at the land—one last time for me—and I read aloud from the Bhagavad—Gita, a passage read at Ella Chambers' memorial: "...end and beginning are dreams..."

At Twin Creeks Ranch, our dreams came true.

\* \* \* \* \*

## CHAPTER 29

# Leaving

*"Finis coronat opus [the end crowns the work]."*
—Anonymous Latin saying

CARL STAYED AT THE RANCH FOR THREE MONTHS, fixing it up for show and sale. He was determined to sell it himself, and ultimately he did, to a couple who then were just about the same age as we were when we bought the ranch, 12 years earlier.

They had a 14 year-old boy and a couple of grown children, and seemed as enthusiastic as we had been at first to roll up their sleeves and dig themselves into the ground just as we had done there.

First, they enlarged the vegetable garden and planted a few more fruit trees in the orchard. Next, they started in on the house and made elaborate plans, just as we had, to remodel the studio.

Carl liked them immediately.

Of course, one thing they didn't have to contend with was cows. They made it clear from the beginning that they did not want our cows as part of the deal. They didn't want *any* cows. Boy, were they smart. I was impressed.

The good news is that they're still very happy with their ranch, and have told us we're welcome to visit sometime.

I'm not sure I could ever go back, though. When I said goodbye to the ranch for the last time, it was such an emotional experience, I'm not sure I want to go through that again. It was a place we'd loved, one we'd found together, moved into, and struggled with together. A place where we'd been happy together.

Best leave it in memory. As the Buddha said, you can't step into the same river twice.

\* \* \*

Carl had more than a few distinctive challenges as he prepared the ranch for sale.

Some examples culled from his journal:

Zenia, July 15th, 2002
Seems like I've been up here for two months—maybe only one-and-a-half. I've lost track. First had to fix things up: cut grass, trim underbrush and trees, fix water system—wood rat with huge nest on top of spring box. Dead bull in big slide area. My friend Fernando came up one weekend to shoot deer. Near miss. Fernando by himself for hours. Waiting for deer. Deer appeared right in back of him. He turned and reached for bow and arrow at its feet. Shot arrow and string broke on his bow. Arrow fell at deer's feet. Deer snorted disgustedly and ran away.

Sprayed roses at front door with Liquid Fence. Turned out to be liquid cesspool. Fernan' and I had to run in the house quickly because of the stink. Smell came right through the walls. What would I tell prospective buyers when they came to the door? Smelled like our septic tank was broken.

Ads in paper got three prospects up here—first a mother and father with 13 year-old kid looking for a place for other son to do a vegetable farm. They didn't—couldn't—physically see most of place.

Hugged me goodbye.

Second couple from Petaluma wanted to ride Arabians on trails.

Went to Ella's funeral. She died about two weeks before the 4th. Called her own death, by telling them to pull the plug on the respirator. Ralph all alone now.

Albert took me to see his 160-acre project beyond Double Gate Road in the National Forest. Three ponds in and one more to go. He driving me down cliffs, along the edge of the ridge, along single lane dirt roads in his 4 x 4; holding out his divining rod with one hand, holding the wheel with the other as we bumped along and he tested for underground water.

Albert is our resident dowser. He rarely misses. You want a well but you don't know where to dig? You call Albert.

I have packed two truckloads of stuff out of here and am starting on third. Books last time— weighed a ton.

Zenia, July 16th

Spent a relaxed, aimless morning. Slept late. Hearing a World War II vet talking about his experiences on the radio last night, after which they played "I'll be seeing you in all the old familiar places, that this heart of mine embraces..." I was overcome with emotion and started a like— sounding poem for ML.

Later, Snaz and I went up the hill to hook up the overflow from the highest tank to the orchard. Much slipping on steep sides of the field in non-grip boots on top of Madrone leaves (worse than banana peels). Snazzy gave up on me and went home. Later on, a coyote started yipping and running down the mountain in the forest. I told Snaz: "Go get 'em." She was below me on the road. Instead she came running up the hill to me. She somehow knew better than to give chase.

M.L. called late, after her Friends of the Library picnic and her Mac meeting. Worried about deal going through. Hiked over to riding ring.

Found some star thistle I missed.

Got final offer from buyers. Theirs was $2,000 lower than mine, with furniture, so we flipped a coin for the $2,000. I lost.

Zenia, July 19th

Today is Friday, I think. Got up after feeding Snaz and then going back to bed. Spent the whole day packing. After supper, cleaned out stock water tank and put more 9' peelers in the truck bed.

As if the nasty smell of the roses by the door wasn't bad enough, another odd thing happened as I was cleaning house in anticipation of arrival of prospective buyers:

Was carefully vacuuming floor, when something fell through a space in the tongue-and-groove ceiling and landed at my feet. Something squirmy.

Looking closer, I realized they were maggots!

Some animal had died in the attic crawlspace, and they were voraciously cleaning it up. Had to busily clean *them* up before the folks arrived.

The final indignity.

\* \* \*

Another thing we had to do before selling the ranch

(and before our old bridge collapsed completely into the creek bed, taking us or a few of our heavier cows with it) was to buy, haul and install a new bridge over Burgess Creek.

Carl had been shoring-up the old wooden railroad flatcar that was our bridge for at least a decade, and the time had definitely come to replace it.

Carl Jacking Bridge Up—Again

Fatigue and termites had set in, and the bridge had nearly big enough holes in it to swallow a small calf or two.

Carl hemmed and hawed about the problem for quite a while, but he finally started the search. That was the easy part.

Once purchased, the bridge had to be hauled to the ranch by a very big truck, and once there, it had to be hauled off the truck bed and maneuvered down the hill and hauled across the stream.

Our Gorgeous New Bridge over Burgess Creek

This meant that a very large piece of equipment had to ford the stream, so the new bridge could be winched across, and the old bridge destroyed underneath. Poor guy had to go way downhill, so he could cross where the creek bed was shallow, and come back to the other side of the old bridge, where we waited.

Then came the delicate job of getting the new bridge across quickly, before the old one collapsed under it.

It was hairy, and not without some head scratching. But, they did it—two heavy equipment guys and Carl, in the midst of it all as usual, shouting directions.

When it was in place, we breathed a collective sigh of relief.

"And now," said Carl, as he picked up his checkbook and brandished his ball point pen when all was finished, "for the *really* heavy lifting!"

\* \* \*

The time I went up to spend a week at the ranch just before we had to leave it forever, I was armed with our new digital camera, and the knowledge I had gleaned from reading the handbook through the week before. I went to make new memories and recapture old ones.

So many memories.

The time we discovered we had our very own bog, and got stuck in it up to our boot-tops.

The night of the many-colored shooting stars.

The times we watched bats fly over our lake at dusk. So many memories.

The lake one summer, when Carl made a makeshift dock out of thick foam insulating material, and we used it to launch ourselves into the deepest water, to swim and paddle—Snazzy always right behind us, trying hard to paddle onto our backs so she could use us as a flotation device, blowing bubbles, trying to bite the water; she was happy—so were we.

Our memories still have the power to warm us or bring us to tears.

So many memories.

Carl snorkeling around the lake, diving down, trying to find the springs that fed it. The coolness of it on hot summer days; many hot summer days.

Discovering wild ducks swimming in the lake in the cool early morning; watching the large-mouth bass jumping for dragonflies on warm evenings; and one morning when we managed to hold the dog back enough to sneak up on the lake and see what we might see.

We were not disappointed. There, swimming lazily and happily in the middle of our pond was a river otter, a big one. At first he didn't see us, and when he turned his large slick head lazily in our direction, it took him a couple of seconds to figure out just what he was seeing.

Then he did something so comical I couldn't believe

what *I* was seeing. It was a classic double-take. As he stared at us the second time, comprehension dawned, and before we could let out a breath he was gone, leaving nothing but large ripples behind.

We turned to each other with astonished joy. "It was an otter," Carl gasped, stating the obvious.

"I can't believe it," I said. "We actually got to see one, and he was right here in our own pond! Hallelujah!"

So many memories.

That spring I found and identified and cataloged 85 different varieties of wildflowers growing on our land: the Indian pinks, the five spot, the elegant cat's ears, the wild iris, the trilliums, the camas lilies, the dog-toothed violets, the now endangered Douglas' meadowfoam.

I'll never forget the wonderful people who helped us learn the ropes.

Nor will I soon forget hearing the strange sounds in the night—sounds that froze the blood. Coyotes? Mountain lions? Yetis? I'll never know.

Or at the Teapot Dome sitting, looking out over the valley, watching the deer come out to feed at dusk, and the bats come out to fly.

Or the times we sat cooling our feet in the icy stream below the big waterfalls on Burgess Creek, watching small trout swimming in the deep pools below us and listening to the amazing thundering of the falls, foaming madly in the deep green beginning of each passing year.

And still the ranch did not want to let us go.

\* \* \* \* \*

## CHAPTER 30

# Paradise Lost

*"If there is a paradise on the face of the earth,*
*It is this, oh! It is this, oh! It is this."*
—Mogul Inscription in the Red Fort at Delhi

Our bucolic odyssey had come to an end, amid tears of both sadness and relief.

It was good to concentrate on just one garden for a change; just one house, one set of repairs, one large pet.

Do we miss it? Yes, and no.

Certainly we do not miss the four-hour commute each way.

We don't miss the frozen, broken water lines, the over three miles of cattle fencing that needed constant patching and replacing, the iffy water supply, the ankle-turning holes in the beautiful meadow, the leaky roof, the fallen trees that needed to be constantly sawn up and removed from roads and streams, the mud, the landslides, the star thistle, the mice in the walls, the rattlesnakes under the house, the raccoons in the attic, the mud-dauber wasps in the bedroom.

But, of course, we don't think of those things anymore.

Instead, we recall the depth of our learning about ourselves and about each other. We made the happy discovery during our 12 years at the ranch that we were more multitalented and resilient than we had realized, and we honored that, in ourselves and each other. We understood, finally, the amazing range of Nature and our place in it, which, it turns out, is pretty darned tiny; about the vastness of the universe and the workings of rust and mold, and of the tiny creatures who inhabit the microcosm of the garden and make things work.

Instead, when we close our eyes we see our lake: covered with mist and wild ducks at dawn; with dragonflies and damselflies and leaping fish in the still, somnolent summer afternoon; with swooping bats and swallows at sunset.

Carl no longer remembers the bodily pain or the exhausting labor we contributed to the ranch and its inhabitants, nor the visiting bull that charged him. (Carl decided then that the best defense was an offense, if he was to avoid being gored (*that* one had horns).

Now, he thinks often and fondly of the cows, with their amazing and distinct personalities.

He thinks of how he used to go down to the pasture on warm summer nights and sit with them, absorbing their quiet dignity and inscrutable beings.

I think now of the sweetness of each returning spring, which brought with it the myriad of wildflowers, some of which I had never seen before; the feeling that we were not so much stewards of the land, but rather that the land would benevolently tolerate us, for a time.

It was humbling.

We both remember with fondness the ritual we practiced together each time we left the ranch.

We'd sit on a bench overlooking the cattle barns and the big meadow, or stop in the driveway in the pickup once we were loaded up for the four-hour trip back to town, and have a moment of silence and centering.

Then, we'd tell each other what we remembered best about that most recent stay at the ranch—the good and the bad—and what we had learned from it.

We'd think about how everything always went on pretty well without us—the cows, the cats, the flowers and trees, the bugs and the ferns and the lichen-covered rocks and the deer and the small, furry animals, the creeks and the lakes and the ducks and the fish.

And how Nature had the last word, always.

As we pulled away (why was the weather always so beautiful when we got ready to leave?), we would say farewell. I would say, "Goodbye, Twin Creeks. Be safe, and keep the animals safe while we're gone."

The feeling was like saying goodbye to a child as he or she embarked on a life that no longer included you, that would never include you again, except on rare occasions—the same almost physical tug at the heart, at the emotions.

Then we would stare off across the majestic vastness of land, trees, mountains, and sky, and I would whisper:

"We love you, Twin Creeks."

\* \* \*

We still see the rumbling waterfalls, and the fog that hovered above the valley floor over Alderpoint, turning the edge of our big meadow into a lake of clouds.

We still see the pink beginning of new oak leaves and the green promise of wild grasses rippling in the wind like ocean waves; the daffodils and tulips and iris I planted, each in turn exploding with bloom like fireworks hanging in mid-air; the mossy, leafy banks of the streams as they rush and murmur throughout the seasons; the silence of the winter snow as it fell, transforming the ground on sunny days to trillions of tiny reflecting jewels.

And often we recall the early evenings talking by the fire; the hot tea brewed on our Franklin woodstove; the glass of wine reflecting the candles burning on the dining table after a fresh, home-grown meal; the dog curled up just beyond the door and the cats curled up on the roof, near the chimney.

The balmy summer evenings on the deck, watching the sunset over the King Range; craning our necks skyward to marvel at the stars; the enveloping peace of the place...

...The feeling of being home.

\* \* \* \* \*

# Ranch Poetry

### ZENIA CHRISTMAS HAIKU

Snow on the mountains
Calves around the barn
Loved ones close at hand
No tenants on my answering machine
These are a few of my favorite things.

—Carl R. Sutter


### THE GREAT LIGHT DAWNS

Where have I been that I missed the whole
Consciousness thing?
Was I dozing at a seminar, and failed to
Hear about it?
Being left out at parties is simply not
My style—
I don't mind telling you, I'm pissed. Who the hell has the
Signup sheet?

—Mary Lynn Archibald


### CLOUDS

Coastal clouds; dark thunderheads,
Puffy cumulonimbus, threatening, gray,
Amorphous rain clouds,
Mare's-tails, ribbons, layers, whorls,

*221*

Pink, purple, white and gray,
Foggy wisps,
Mountain-girdling phantoms,
Melting.

—Mary Lynn Archibald

Sun

Streaming, flashing, blessing,
Warming the cold, damp earth.
Caressing the trees and coaxing the plants
To swell with new growth; new life.
Revealing the valley in all its new green mantle,
Like a velvet gown; the grass so delicate,
High and lush.

—Mary Lynn Archibald

Love Poem, for Carl

I turned toward you like a
Blind, pink piglet
Knowing it is about to
Find the teat.

Your body knowing mine
As sure as if
Our history was composed
Of millennia, not years.

You held me and it was
As if you felt as much—

I dared to hope you, too,
Were coming home.


PARTLY CLOUDY

When the dark fog lifts
I am awestruck—
Prostrate at the Golden feet of the
Unknowable.

So many things stay
Unnoticed in the
Rush of what is lightly known as
Living—

We fail to see the
Beauty that surrounds us—
Shimmering; beckoning us ever surely
Home.

—Mary Lynn Archibald


STEALING HOME

I love you fiercely—
As you rise to meet my silence
Like the rising storm clouds that
Gather in the west.

I love you softly—
As night wind whispers in dark trees
Gentle as fog that wraps us
In its silken robes.

I love you hotly—
Locking legs and crushing breasts;
White fire of driving need
Touches heart's hub.

I love you sweetly—
Turning conscious thought to wonder:
Rushing headlong, wordlessly
Toward your deep soul.

—Mary Lynn Archibald
2002

SNAZZY

When we sold the cows
The ranch was never the same.
When we put you down, Snaz,
Our lives were never the same.
Now the meadow is empty
Of energetic greetings.
No tail-wagging assaults
At the front door.

Gone are your languid glances
From earthy resting places
As we ant-like humans
Bustled about mysterious objects.
I miss your regulatory movements
From sunlight to shade
For body temperature control.

You never returned the stick—
A small grasp on doggie freedom,
And how slowly you came

When I clasped on the leash
Of mistrust.

Yet in the end
You maintained
Adult species dignity
As you stared off to the side,
Your head held horizontal
To the earth's plane.
The poise of a Grand Dame
That graciously tolerated
Our demands and caprices.

What was going on behind
Those deep-set dark eyes?
Different pupils from mine—
Another species difference,
So much of you a mystery.

Protector
Compaion
Mother
Forgiving in an instant,
Too full of exuberant life
To hold a grudge.

I got in on a few dog secrets—
The sniffings of the air
In rapid succession
To put you on alert;
You looking back for confirmation
When you challenged an intruder.

The drooping of your ears
When you saw me coming;

Deferring to Top Dog.
We never got past that,
Not even with a piece of meat,
Or a tempting throw-stick.
I was ashamed that you feared me.

In the end,
Minutes before your death,
You did not drop your ears;
I sensed our inter-species connection
Dissolving.
Your gaze on something
A long way from us,
Toward the other side of the lake—
(always a place of intense interest).
Your large head immobile,
Wedged between death and life.
Before the final indignities—
The shaving of your leg,
The insertion of the needle.

Your great head fell
Lifeless
Upon your outstretched paws,
And our hearts smashed
Against the disbelief
Of your leaving us.

—Carl R. Sutter

\* \* \*

ABOUT THE AUTHOR

Mary Lynn Archibald is a freelance copywriter and author of two memoirs: *Briarhopper*, (a woman's odyssey from Kentucky's coal country to California, from her birth in 1913 to 1945 at the end of World War II); and the book you now hold in your hands (you lucky devil!), *Accidental Cowgirl: Six Cows, No Horse and No Clue*.

Her online credits include: *What To Do When Your Computer Bites the Hand That Feeds It*, *Staging Your Home for Quick Sale*, and *My Life As A Ghost*.

She has also been published both nationally and internationally, in *Chicken Soup for the Single Parent's Soul*; and in magazines, where her articles have appeared on such subjects as interior design and remodeling, growing lavender, antique roses, wildflower meadows, and producing your own olive oil.

Mary Lynn is a regular contributor to *New York Times* affiliate, Santa Rosa, California's *The Press Democrat*, where her articles have appeared since 2003.

Her writing specialties include business communications (especially in the fields of building, real estate, gardening, and interior design), as well as small-scale commercial gardening, farming and ranching.

She also authors a free, frequently published newsletter on the subjects of country living, gardening, low-cost decorating, small business, and sustainable agriculture, available by subscription at her website:

www.winecountrywriter.com, or
e- mail: marylynn@winecountrywriter.com.

Mary Lynn is a prolific writer; avid reader; entertaining speaker, and mother of two grown children. She is that rarest of creatures, a Native Californian, (born in the tiny town of Soquel, just outside Santa Cruz), and her varied background includes such exotic professions as switchboard operator; art and English teacher; ballerina; interior designer; sales clerk; herb grower, fashion model and chorus girl (not necessarily in that order).

Despite the glamorous pull of the city, Mary Lynn has always been a country girl at heart—never happier than when she has dirt under her fingernails. She and her partner-in-crime, Carl, and their faithful dog, Fizzbo, now make their home in the country, a few hours north of San Francisco.

# Order Form

You may have additional, autographed copies of
*Accidental Cowgirl*, or
*Briarhopper: A History*,
please check here and return to the address below:

Accidental Cowgirl   Quantity_____

Briarhopper             Quantity_____

PLEASE NOTE: 10% discount on orders of 10 or
more. 5% discount additional for book clubs, on
orders of 10 or more.

Please make checks payable to :

ML Archibald
Wine Country Writer
1083 Vine Street, #185
Healdsburg, CA 95448
Phone/Fax: (707) 395-0542